Discover the Relation Between Fashion, Fabrics, Weather, and Comfort

Nicole Mölders

DEDICATION

This book is dedicated to my husband, Prof. Dr. Gerhard Kramm.

ACKNOWLEDGEMENTS

I would like to thank Profs. Drs. Gerhard Kramm and Joan Braddock for encouraging me to write this book, and Prof. Dr. Kramm for fruitful discussion and helpful comments. Furthermore, my thanks go to the team from Amazon Publishing Pros for help with the layout and finding the previous-last typos.

ABOUT THE AUTHOR

Nicole Mölders earned her BS and Diploma in meteorology, her doctorate in Geophysics at the University of Cologne, and her habilitation in Meteorology at the University of Leipzig in 1983, 1988, 1992, and 1999, respectively. As a scientist, she published various peer-reviewed papers related to weather, environmental sciences, climate, radiation, and thermal comfort. Furthermore, she is the author of several atmospheric sciences and fashion books. In Germany, at the University of Leipzig, and the US, at the University of Alaska Fairbanks, she taught various classes in face-to-face, hybrid, and online mode.

Today, she is an Emerita Professor of Atmospheric Sciences, scholar, author, and a fashion and lifestyle blogger from High Latitude Style (https://highlatitudestyle.com). She lives with her husband Gerhard Kramm and two cats BT and Isabella in College, Alaska.

Table of Contents

PROLOGUE

This book is for everyone interested and/or engaged in the fields of textiles, fashion, fashion design, thermal comfort, and/or weather, especially meteorologists, bioclimatologists, engineers, technologists of the textile industry, apparel salespersons, fashion designers, high school students interested in the STEM field, and laymen. The book can also serve as a textbook for a 100-level science class for non-science majors, especially for fashion design and fashion majors, in community, technical, and fashion design colleges, as well as universities.

This book combines my love for clothes, energetics, and weather. One of this book's goals is to inspire high school seniors and freshmen students to pursue a career in fashion or textile engineering. Another goal is educating fashion lovers (in an entertaining way) to create a wardrobe that works for the weather of their climate zone. Goosebumps and sunburns are never in style! Nobody looks their best when they feel thermal discomfort — whether it is due to cold or heat stress.

The book swaps out myths like, for instance, "The coat warms." From an energetic point of view, the statement is wrong. It would violate the *Second Law of Thermodynamics*. Think about it! Your reaction will be, "Of course!" If the coat could really warm something, it would warm the room, too, when it hangs on the coat rack. But, of course, the coat doesn't do that! Any warming needs an energy source like the little batteries in your heating gloves or the electricity that powers your grandma's heating blanket. This means the coat only warms you when you burn it! Therefore, correct statements would be, "The coat keeps you warm" or "The coat provides good insulation."

The focus is on the energetics of clothes related to thermo-physiological comfort in various weather conditions. After reading this book, people can judge the thermal comfort of garments when reading the clothes tags. The book excludes physical aspects related to quality and durability, like yarn strength, breakability, abrasion, and tendency to pill or fade. These quantities vary strongly but are irrelevant to dress weather-appropriately. Furthermore, clothes tags, unfortunately, do not disclose these quantities.

1 THE ROLE OF CLOTHING FOR HUMANKIND

Clothes have played a key role in the survival of humankind since ancient times. They namely protect people from the elements of weather and other harmful dangers of their environments. Although the human body can perform auto-thermal-regulation via shivering (muscle movement), sweating, arrowing (vasoconstriction), and/or widening of blood vessels (vasodilation), the human skin is vulnerable to harsh environmental conditions. Intense sunlight, rain, snow, and strong winds, particularly, can change skin temperature immediately, causing discomfort, thermal stress, heatstroke, hypothermia, or even death.

Despite the fact that modern human has been educated in natural sciences within the K-12 framework, today, heat and cold-related losses of life are high [1], [2], [3]. A major reason is that, in today's modern society, we often take thermal comfort for granted — at least indoors. Here thermal comfort means that 1) we experience no thermoregulatory sweating, 2) our only moisture loss is through the skin and breathing, and 3) ambient temperatures are between 48.2 and 71.6F (9-22°C). Just think of the benefits of air conditioning (AC) during summer or heating in winter. But many old buildings lack AC, and daily activities often require people to step outside. Once outdoors, we are exposed to and have to protect ourselves from the elements of the local weather.

Sure, urban planners advocate for shaded areas, parks and urban forests, and green or white roofs to reduce the urban heat island effect [4]. Such measures can lower air temperature and, hence, improve thermal comfort in their immediate vicinity and slightly downwind. However, these measures cannot reduce high relative humidity.

Our bodies react with transpiration, shivering, reduced blood flow, etc., to cope with thermal heat and cold stress, respectively. However, there are metabolic and physical limits to these auto-regulatory functions. For instance, high relative humidity in combination with heat may suppress the physical processes needed to create the beneficial cooling of the body from transpiration. Consequently, each year many people suffer from the summer heat and heatwaves. Signs of thermal heat stress are having trouble sleeping and feeling tired, hot, or stressed, just to mention a few. During times with such weather conditions, mortality rates go up [1], [2], [3]. Can clothing help to reduce thermal heat stress? If so, what physical properties do the fabrics need to have? Does wearing white during summer help, as it is commonly believed?

Unfortunately, our body's reaction to cold stress fails to protect us from hypothermia caused by long-term exposure to below-freezing temperatures. Actually, percentage-wise, cold weather causes more deaths than hot weather. The after-effects of exposure to low temperatures (e.g., cold, flu, pneumonia) all contribute to the number of deaths related to cold stress. However, when looking at absolute numbers, more deaths occur due to hot than cold weather in developed countries. The reason is that fewer people live in extremely cold regions. Furthermore, people living in cold regions tend to avoid spending time outside when it is extremely cold. Or, at least, they limit their time outside to the absolute necessary, if possible.

However, there are people who have to work outside for an extended amount of time, even when it is extremely cold. This group includes soldiers deployed into the Arctic or high mountainous terrain, scientists performing polar research in the Arctic or Antarctica, workers on Arctic oil platforms, and all kind of athletes involved in winter sports, including mushers and snow machine racers. So how do they survive?

Another thread is long exposure to the sun, which is a major cause of skin cancer [5]. Sunburns are common both in hot and cold weather. How can you avoid long sun exposure? Does staying in the shade help?

1.1 Brief History of the Development of Clothing

Clothing distinguishes humankind from other living beings. Clothing shields our skin from the environment. Obviously, clothing needs differ for the various weather conditions, climate zones, and environments. Nevertheless, the developed clothes commonly used energetical concepts for protection from the elements. Even though the early *homo sapiens* had no physics education, they understood that clothes could 1) protect their skin from harm by sunburn, frostbite, and injury and 2) increase the duration of feeling comfortable in windy, rainy, and cold weather.

Let's go back in time for a moment to see what our ancestors already discovered. The oldest clothes found date back to about 7500 BC in the Middle East. Although we have no clothes remains from before 7500 BC, many indirect signs of fitted clothing exist. Archeologists found primitive hide scrapes, cutting devices, and needles (Fig. 1.1). The few finds of preserved clothing, fabric fibers, fur, and leather provide insights into the clothes of ancient societies. Across human history, the development of clothing and textiles depended on the environmental conditions, local weather, progress in civilization, and the technologies to make the garments. Therefore, items needed for clothes production permit conclusions on the kind of clothing worn at a time in the past and sometimes even on the societal structure. Gilligan [6], for instance, interprets the invention of the first eyed needles (e.g., Fig. 1.1) as the beginning of the use of underwear or undergarments. Holed clothing, namely, requires no sewing.

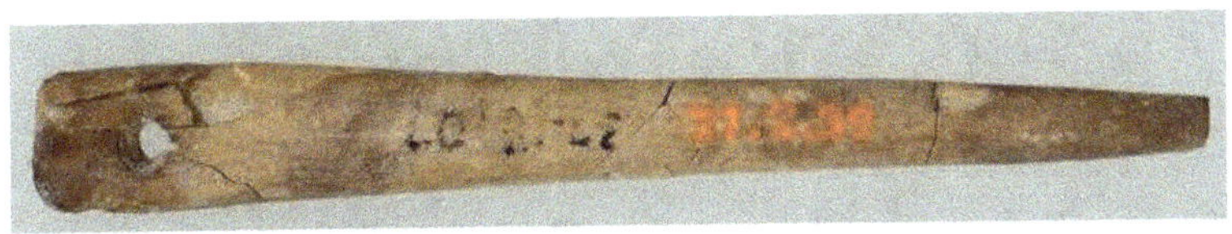

Fig. 1.1. Ivory needle with eye. Predynastic, Badarian ca. 4400–3800 BC. From Northern Upper Egypt, Matmar, Tomb 3107, BSAE/Brunton excavations, 1931. 8 cm (3 1/8 in). Rogers Fund, 1932. Metropolitan Museum. Public domain.

The creation of clothes for thermo-physiological comfort was a long road of optimizing, paved with trial, error, and learning. When, for instance, during the *Ice Age*, humankind started using the skins of the animals they ate, their goal was improved thermal comfort. These humans found out that leaving the hair on the skin kept their bodies warm for a longer time than scrapping them off. Consequently, they started using bare hides and fur in the warm and cold seasons, respectively. What is the physics behind the better insulation of fur than bare hides? Simple: lots of thin, still air pockets, which provide good insulation. That's why faux fur pops up as a trend in fashion every now and then.

Going forward in time, humankind developed techniques to create felt from wool. The oldest felt finds date back to 6500 BC in Turkey. To produce felt, the *Neolithic* people of Central Asia and the Near East combed wool. Thereafter, they laid the wet wool on mats that they rolled tightly. Heavy sticks served to beaten the mat. Due to this procedure, the strands of wool clung to one another. The result was a well-insulating, durable, pliable fabric, which they used as rugs or cut to sew clothing and tents. The purposes of the garments, tents, and rugs were to mitigate the thermal discomfort due to the cold air and ground. The major advantages of felt over hide/fur are the lower weight, the control of the measures, and thickness. In addition, scraped-off wool was no longer a waste.

Humankind has used various plant materials and fibers to create fabrics. Many archaeological textiles apparently consist of flax (*Linumusitatissimum*). According to archeological data, hemp has been cultivated and used already around 6000 BC. The *Er Ya*, the earliest Chinese dictionary, written about 2200 years ago, documents that the Chinese recognized the sex of the plants: "Male hemp is called *xi ma*, female hemp is called *ju ma*." It also documents the use of thin, soft, male hemp for clothing [7].

Paleontological finds of pollen and seeds dating back to 350 BC–450 AD evidence hemp and flax cultivation in northern Scandinavia. However, these data fail to prove the textile use of flax and hemp because the seeds can also provide oil and food.

A tapestry in a Norwegian ship dated 800 AD showed, besides wool fibers, decayed remains of hemp and flax. Archeological finds from *medieval times* showed that the Vikings used hemp and flax in blended and fine household textiles as well as for ropes and sail cloths [8].

Both hemp and flax grow well on soils suitable for crops. However, in contrast to many other crops, hemp also grows on muddy soils and soils with high pH values without special agricultural techniques like burn-beating, plowing, crop rotation, or lie fallow. Muddy or high pH soils are common in mountainous regions, coniferous forests, swamps, marshes, and around lakes like they exist, for instance, in Scandinavia, Siberia, and Alaska.

The ancient humans living in hot climates searched for protection from heat and the sun. In the *Early Dynastic Period* (3150–2686 BC), for instance, the ancient Egyptians shaved their heads. They protected their blade heads from sunburn with hats. Wigs and elaborate headdresses became popular among both genders of the upper class (Fig. 1.2). Despite serving the same primary purpose, the head cover distinguished the classes depending on the societal status and income. Why was the hat, wig, or headdress better than the own hair? Again, physics. The tight material better reduced the penetration of sunrays to the scalp than the own hair. But what characteristics do fabrics possess to be able to do so?

Fig. 1.2. Standing woman with wig. Early Dynastic Period ca. 3100–2649 BC. From Northern Upper Egypt, Abydos, Osiris Temple, Chamber M69. Ivory. 4.6×1.9×1.6 cm (1 13/16×3/4×5/8 in). Gift of Egypt Exploration Fund, 1904. Metropolitan Museum. Public Domain.

Over time, *homo sapiens* developed various methods to improve the desired comfort properties of fabric. Roughening a fabric, for instance, creates little air pockets that increase insulation as compared to the smooth structure. Flannel made of heavy cotton or cotton blends is an example of such treatment with a twilled face and a napped back. Due to its strength, "warmth," and absorbance, flannel is great for interlinings, sleepwear, and bedding in cold climate regions.

Another method to increase insulation involved cooking and brushing the fabric (e.g., *loden, melton*). To produce loden, a loose weave of strong yarns spun from the coarse oily wool of mountain sheep undergoes a repeated process of shrinking due to hot water, brushing, and clipping of the nap. Brushing of cloth increases the teasel and gives the fabric a dense, felt-like texture. The process is repeated until the fabric provides good insulation for its areal weight and becomes wind-proof and very durable. Melton, the fabric of the original Duffle Coats, is made in a similar way. In the production of loden, melton, and flannel, our ancestors used their experience that thermal insulation increases with decreasing fabric density.

Let's make a time jump. The inventions of industrial looms, sewing, spinning, and knitting machines permitted the production of more clothing. Crafting clothes went from individual custom-tailored or upcycled items to mass production. Because of an overabundance of cheap, often poor/low-quality clothes, clothing became more affordable. The middle class could now buy clothing beyond their everyday needs. Prior to these inventions, only the nobility and the rich could afford large wardrobes with more than one new outfit per season.

1.1.1 Clothing Invented for Cold Snow Climates

The Arctic aboriginal people developed insulated boots called *Mukluks*[1] to cope with their cold weather. The word "mukluk" is of Yupik origin and derived from *maklak,* which means bearded seal. Traditionally, mukluks are soft boots with a flexible sole made of sealskin and often decorated with beading. The original mukluks belong to the best insulating winter boots designed for extreme temperature and snow conditions. Their material permits air exchange and the feet to breathe. Breathability is advantageous in extremely cold conditions because transpiration may contribute to causing frostbite.

Norwegian Arctic explorers developed the *lobben boots*. Originally, lobben boots came from Nesna, Norway. But today, a small family-owned company produces them in Tartu, Estonia. Lobben boots consist of boiled, pressed, felted wool with a ¼ inch (0.6 cm) thick wool felt insole. Therefore, they are light. The wool felt provides excellent insulation and comfort between -49 and 23F (-45 to -5°C). Lobben boots also may have a cotton lining. Because boiled wool breathes, your feet stay dry, even when they are perspiring. The natural lanolin of the wool repels (melt-)water. The polyurethane soles' deeply ridged profile provides excellent traction on snow and ice. Lobben

---

[1] Mukluks for sale to tourists are often made from moose- or caribou-hide. However, exchanging materials alters the physical properties of the item, and may impair the suitability for the original purpose. The industrial manufacturers of cold-weather boots adopted the design of the mukluk. However, they use a rugged contemporary sole, cow-hide, and often add faux fur trim. Due to these changes the industrial mukluk provides less insulation than the original.

boots are unsuitable in cold rainy weather or in the melting season, when puddles are all over the place.

The upper part of the lobben boots can be round or sewn like a moccasin. When you are likely to walk in deep snow a lot, opt for the round top. The moccasin design collects snow that withdraws heat from your feet. However, they are the more stylish choice when you only walk on cleared ways. Lobben boots look great in casual outfits with jeans. In recent years, the popularity of lobben boots has increased among Alaskans[2] because they are the first choice of many Iditarod and Yukon Quest mushers who have celebrity status.

Extreme cold vapor barrier boots — so-called *bunny boots*[3] — were originally designed for the military. The Navy Clothing and Textile Research Center in Natick, MA, USA developed them for use during the Korean War. These boots and their cousin, the *Mickey Mouse boots*, are a staple cold weather gear in civilian work and recreational environments in winter in cold climate regions. The insulation of these bulbous, bulky boots works by sandwiching up to one inch (2.54 cm) of wool and felt between two layers of rubber. These boots have a pressure-release valve. Therefore, they are a favorite among bush pilots. Because bunny boots are relatively heavy and clunky, they are not great for extended walks. Typically, they are worn with heavy wool socks.

A (shapka) *ushanka* (Russian for ear-hat), aka *trapper hat*, is a cap with ear flaps. The ear flaps can be tied at the crown to stay out of the way, tied in the back to cover the ears only, or tied under the chin to keep the ears and chin warm. The ancient *Scythians* and various nomadic people in Central Asia and the Arctic used such hats too. During the Russian Civil War, a version with an additional eye flap was introduced as winter uniform. This version became a symbol of the former Soviet Union. During his 1974 visit to the Soviet Union, US President Gerald Ford wore such a cap (Fig. 1.3).

Historically, the trapper hat was made of sheepskin, rabbit, muskrat, mink, or sable to protect the head and ears from cold air and the head when slipping on the ice. Mink keeps the ears and chin safe even at -94F (-70°C). Today trapper hats often consist of faux fur (aka

---

[2] Scandinavians have worn lobben boots since 1955.

[3] Bunny boots are not manufactured anymore, but you can find them used at Army/Navy surplus stores, on eBay or Craig's list.

fish fur[4]). The simplest version of faux fur is of wool pile with a cloth substrate and cloth top, except for the flaps where the pile is exposed.

Fig. 1.3. Photo of President Gerald Ford with Soviet General Secretary Leonid Brezhnev upon Ford's arrival at Vozdvizhenka Airport, Vladivostok, U.S.S.R. 11-23-1974 by David Hume Kennerly. Gerald R. Ford Presidential Library: A2102-06. Public Domain.

A variation of the trapper hat is the *aviator hat*. Because temperature decreases with height in the troposphere[5], the aviator hat served to keep the pilots' heads and ears warm when they flew at high altitudes[6] with open cockpits. Later when closed cockpits became available, the aviator hat remained popular among bush pilots even in summer due to the unheated small planes. Aviator hats were extremely popular among boys in the 1950s because the pilots of WWII were their heroes.

Astrakhan caps, aka *Karakul, Canadian wedges* or *ambassadors,* are made from the pelt of a young Karakul lamb. Note that the lamb is harvested immediately after, or even shortly before, birth when the hair is still black, soft, and tightly curled. This procedure also means the death of the mother sheep. The style of the hat differs among cultures. Most known is the flat, round-top version with double brim worn by the

---

[4] The term fish fur refers to the fact that the fur has no relation to any real fur.

[5] Troposphere refers to the first 6.2 miles (10 km) or so of the atmosphere.

[6] Temperature decreases about 11.7F (6.5°C) per 1094 yards (1000 m) of height gain.

Soviet Politburo members[7] (see Fig. 1.3). In northern India and Pakistan, a slightly higher version with a single crown — called *Jinnah cap* after the founder of Pakistan — is popular. In Afghanistan, a triangular version has been worn for centuries. The winter hats of the Royal Canadian Mounted Police are often incorrectly called *astrakhan*.

1.1.2 Garments Developed for Humid, Rainy, Temperate Weather

The *janker* or *spencer* is a boxy hip-length coarse loden jacket or coat originally worn in the Alpine region of Austria, Bavaria, South Tyrol[8], and Switzerland. Loden has a traditional bluish-green color. The name of the fabric goes back to the Middle High German word *lode* or the Old High German word *lodo*, which mean coarse cloth.

The short, everyday version has a knit trim on the seam, closure, and small pockets. Men's jankers feature dear horn or metal coin-like buttons with eagle imprints, while women's jankers have eagle-featuring (or *edelweiss*-featuring) metal buttons. The knee-length coat version, aka *Innsbrucker,* is great for humid, windy weather at temperatures around the freezing point. The women's version often has a hood. Sometimes, embroidery or a trim along the front closure and/or pockets adore the women's garments. Still today, small, family-owned companies produce these traditional jackets and coats and provide employment to many people in the European Alps.

When you do some research on the pea coat (or pea jacket), you find different origins. Probably, like with many things, people independently of each other had the same idea, or at least quite similar ideas, to solve a problem. In the case of the pea coat, the problem was how to protect a sailor from wind and rain at temperatures down to around the freezing point. Pea coats have existed for at least more than 300 years. The word *pea* most likely stems from the Dutch or West Frisian word *pijjekker.* Herein *pij* refers to the coarse kind of twilled blue fabric with a nap on one side, and *jekker* refers to jacket. The fabric goes back to the 16th century when Dutch commerce had a strong naval component.

In British English, the pea coat is often called a *"reefer coat."* Reefers are sailors who climb up the riggings of sailing ships. The reefer jacket

---

[7] Note that very nice faux fur versions are available if you really like the look of a politburo hat.

[8] Since after the Great War, South Tyrol belongs to Italy.

may also have a sporting origin. American newspapers mention the pea coat as early as the 1720s. A further-developed version is the *bridge coat*. Its length reaches to the tights and was the uniform of naval officers. The extended length permits standing on the open bridge for a long time, even in bad weather. The bridge coat has a half-belt in the back, epaulets indicating the rank, and a double set of golden brass buttons on the front. Other further developments are the double-breasted and single-breasted *Navy blazer*.

Originally, this double-breasted coat was made from navy-colored heavy wool. However, in the 1900s, the cut was commonly used with lighter fabric for men's and boys' suits. Historically, in the US Navy, pea coats were made from 30 oz (approx. 850 g) dark-blue wool, typically heavy melton cloth, until the 1970s. Due to its tightness, heavy melton-wool cloth is durable, provides insulation, and keeps wind and even some sea spray or rain out. Modern (fashion) pea coats typically provide less insulation than the Navy version.

A pea coat is narrow at the waist and flares out over the hip to accommodate climbing. This cut is very flattering for women. The oversized collar can be worn up to protect the sailor or today's wearer from the elements at sea and land alike. Left open, it helps to regulate the body heat when climbing or walking fast. In the front, the two overlapping layers increase insulation and protect the wearer from heat loss due to wind. The lower front lapels are cut at rectangular angles for maximum overlap. In the back, the pea coat has a center dart that is sewn flat to avoid cold gaps. The pea coat is the only coat for which the designer had pockets for cold hands in mind. The front has vertical (slash) pockets for easy access. The right outside pocket has a small change pocket. The pea coat has six or eight large wooden or metal buttons for closure, one button to close the collar at the neck, and one inside button to anchor the double-breasted front. Most likely, the first pea coats had short side vents or no vents at all. Today, the US Navy pea coats have a center vent.

Be aware that every vent means a *cold bridge*. Furthermore, the location of vents and their length easily date your coat. Ladies with a womanly derriere should avoid a center or double vent. A center vent namely might split and look like two large tail feathers; a double vent may look like a sheet of a pad calendar.

The *duffel coat* is another garment designed for cool, humid, wet weather. Its fabric is also a coarse, thick, woolen material called *duffel*.

This name refers to the town of its origin Duffel, Belgium[9]. During the Great War, the British Royal Navy issued a camel-colored duffel coat with a bucket-type hood. This military coat reached the knees for protection from winds and cold air. The original British-style duffle coat had a genuine double-weave duffel with a tartan pattern on the inside and solid camel color on the outside. First, the closure featured three, later four, wooden or buffalo-horn toggles with leather loops.

During World War II, a revised design of the duffle coat got the name *convoy coat* or *Monty coat* because Field Marshal Montgomery wore it as part of his uniform. After WWII, the manufacturer sold their large stocks of surplus duffle coats to the public at affordable prices. Due to their high quality and good weather protection, duffle coats became popular in the 1950s and 1960s. They have remained a classic ever since.

Duffle coats are classic, stylish, and great for humid, windy weather. Their straight cut looks great on men, but leaves something to desire when worn by women. However, for a straight-up-and-down slim, tall gal, a duffle coat is a great choice as a statement.

The trench coat was developed for rain protection. Two brands claim to its invention – *Aquascutum* and *Burberry*. The name Aquascutum combines the Latin words for water (*aqua*) and shield (*scutum*). Aquascutum had provided officers with raincoats made from waterproof cotton since the 1850s. John Emary, the founder of Aquascutum, received a patent for this first water-repellent fabric in 1853. Aquascutum coats were worn during the Crimean War. During the Great War, the company provided officers coats with removable, buttoned-in linings.

Thomas Burberry received a patent on his *gabardine*, a tightly woven twill weave, in 1879. In 1901, he submitted the design of the double-breasted, weather-resistant coat to the British War Office. Later Burberry modified this lightweight but durable gabardine coat with D-rings, and shoulder straps to secure gas masks or whistles. The coat was part of the uniform of the British and French soldiers sitting in the trenches of the front line during the Great War, for which the coat got its name. Many war veterans kept their trench coats because of their durability and great wind and rain protection. Like surplus duffle coats, surplus trench coats were sold to the civilian population after the war.

---

[9] The spelling "duffle" is actually incorrect from the perspective of the historic origin. Note that duffle bags were originally also made from duffel.

Hollywood movies made trench coats fashionable by dressing gangsters, detectives, and cops in trench coats in the 1930s and 1940s. Who doesn't associate a trench coat with Humphrey Bogart as Rick Blaine in the movie "Casablanca" or Audrey Hepburn in "Breakfast at Tiffany's" or Winston Churchill?

The key features of the originally-khaki trench coat were double-breasted with ten buttons on the front plus buttons on the waist and wrist belts, wide lapels, a gun flap, raglan sleeves, and shoulder straps. Having grown up in Europe, I insist that one never ever buckles the belt. You either tie it or nudge it in the back.

Hand-knitting was introduced to Ireland in the 17th century. The fishermen cable-knit sweaters, aka *Aran sweaters,* are handmade on the Aran Islands located west of Ireland off Galway Bay. Reports from 1893 suggest that the only knitted items at that time were socks. Some of the younger fishermen had adopted to wear the fishermen sweaters worn by the British and Scottish fishermen. Around 1900, fishermen's wives knitted the first Aran sweaters. The people of these islands made their living by farming and fishing, which both involved long exposure to wet stormy weather. Wool sweaters from untreated (undyed) yarn still have the natural lanolin oils that make the sweaters very water-repellent for both rain and sea spray.

Typically, Aran sweaters feature up to eight different knit patterns. Some sources state that the patterns are related to the family clans of the wearer, and served to identify the fisherman incase he drowned. These sources suggest that the patterns are handed down from one generation to the next. Other sources interpret the patterns as symbolic. The cable pattern, for instance, symbolizes robes and hints at the profession of fishermen. It also stands for hope and should bring luck and fruitful fishing. The most common pattern is the honeycomb design, which stands for the hardworking bee. The zig-zag pattern is said to remind of the ups and downs of marriage.

I am leaning towards the symbolic interpretation for the following reason. The likelihood a drowned man would strand on an Island where they know his clan's pattern is pretty low, especially given the strong currents around the Aran Islands. Of course, as with many handmade items, insiders can distinguish styles and even identify the designer. When I recall my teenage years, my sister's knitting was always one small pattern all over a sweater, while my patterns only repeated when mirroring the pattern required it.

In the 1920s, the Kennedy brothers wore cable-knit sweaters when playing football. Therefore, these sweaters became popular for boys in the US as *Sunday's Best*. Fishermen sweaters became a men's wardrobe staple in the late 1940s and early 1950s. Via the detour of menswear, cable-knit sweaters also became a favorite in women's wear.

In 1940, *Patons of England* was the first to publish an Aran pattern from a store in Galway. *Vogue* published one in 1956. The growing demand for Aran sweaters as a souvenir for Ireland tourists and as a statement item for modern fashion, knitting by hand and machine has become an economic leg on the Aran Islands besides farming, fishing, and tourism. Today, the famous Aran cable-knit sweaters are also available in cashmere, alpaca, silk, cotton, and other yarns. And, of course, there are many "mock" fishermen sweaters from synthetic yarns that lack the advantages of untreated wool. Recall that producing a garment in a different material may cause the loss of some of the original benefits. This fact is important to consider when dressing for thermal comfort.

The name *Fair Isle* refers to a traditional knitting technique named after the Fair Isle, an island belonging to the Shetland Islands north of Scotland. Fair Isle sweaters, aka *Shetland sweaters,* are knitted in a round with a pattern of different colors. Typically, Fair Isle patterns use about five different colors in total, and only two different colors per row. This technique uses either a two-pointed needle with a flexible plastic middle to hold all the loops, or five double-pointed needles. In the latter technique, four needles hold about one-fourth of the meshes. The fifth serves to knit the next quarter of a round. Both ways limit how much a knitter can knit in a given color. To avoid loose strands, the knitter catches the yarn-not-in-use by the yarn-in-use. Typically, catching of unused strands occurs every 3-5 stitches.

Obviously, this knitting technique is very suitable for creating an extra layer of insulation because five colors require a lot of catching-up strands. Because air exists between the various strands and air is a great insulator, the pullover protects the wearer from body-heat loss around the upper chest. Fair Isle sweaters became very popular when the Prince of Wales (later King Edward VIII) wore a Fair Isle sweater-vests in 1921.

Unfortunately, since the 1990s, vendors and designers have used the term "Fair Isle" very loosely for any stranded color knitting. Concretely speaking, the name only refers to the origin of the knitting technique. It no longer identifies the place of manufacture, nor the production following the traditional technique and pattern. Cheap mock versions of Fair Isle/Shetland sweaters have the pattern stitched onto the knit.

The Norwegian *lusekofte* ("lice jacket") goes back to the 8th century. This sweater has its roots in Setesdal, a region along the Otra River valley in southern Norway. Therefore, it is also called *setesdalgenser*, which means *Setesdal sweater*. In the 8th to 11th century, prior to the Vikings, this region consisted of petty kingdoms leading to the development of local traditions for identification. While complex terrain and many borders meant isolation, difficult travel, and trade, they also preserved traditional culture. Therefore, many different knit patterns exist [9].

Norwegian sweaters and jackets were hand-knitted on round needles or five double-pointed needles in a similar knitting technique to the original Fair Isle sweaters. Consequently, they provide thermal comfort for a longer time than a single-yarn knitted garment would. Norwegian jackets have unique decorative embroidery trim that surrounds the neck, closure, and cuffs. The embroidery differed among regions. Typically, silver-colored hooks sewn on both sides of the jacket serve as closure[10].

Traditionally, these sweaters were men's working clothes. Today all gender wear lice sweaters. Wearing this sweater with a shirt and tie counts as business attire in Norway. In cold climate regions, these garments are office appropriate with wool slacks or skirts in business casual workplaces. When made of wool, the lice jacket is great for rainy, cool/snowy, cold winters.

1.1.3 Attire Developed to Beat the Heat

The first *T-shirt* versions were sort of DIY by laborers. During the heat of summer, the typical undergarment — a one-piece union suit — was plain unsuitable for thermal comfort of miners, industrial and dock workers. Consequently, they cut the one-piece union suits into two pieces to make them separate. The upper part then looked like a capital T. The cutting improved ventilation. Moreover, the gaps between the pants and the "T" allowed for better transport of body heat and sweat away from their bodies.

---

[10] Historically, Norwegian clothes relied on brooches or hooks rather than buttons for closure.

Between the Mexican-American War (1898) and 1913, around the time of the Spanish-American War, the US Navy started to provide T-shirts as standard undershirts. These white jersey cotton shirts with crew-neck and short-sleeves were worn under the uniform. In the 1920s, the shirt also became popular as an undershirt for workers in agriculture and industry. The tough fabric felt cool and provided better comfort on warm and hot days than the DIY from a union suit. By the Great Depression, the T-shirt was the default underwear for workers. Because of the easy-to-clean fabric, moms preferred tees as everyday wear for their sons for chores or playing outside. Between the Great War and WWII, the US Army adopted T-shirts as part of the uniforms for training during hot weather. After WWII, veterans wore T-shirts tucked into their trousers as casual clothes.

In the 1950s, the fashion industry saw a new market — *teenager fashion*. Post-war baby-boomer teenagers opposed the mainstream culture. They wore leather motorcycle jackets, motorcycle boots, jeans, and a white T-shirt. Next, Hollywood took up the trend and dressed their main actors in a tee. Just think Marlon Brando in "The Wild One" or "A Street Car Named Desire." These movies made the garment fashionable as mainstream general-purpose casual clothing worn tucked in for summer.

US soldiers stationed in Europe wore the undershirts under their uniform and plain with jeans on weekends. Inspired by Hollywood movies and the soldiers, young European men started to wear T-shirts. As time progressed, T-shirts became mainstream for men on weekends.

In the late 1960s, the T-shirt became shorter on the sides than on the front and back. The *Flower Power* youth wore it untucked and also introduced the T-shirt into the fashion of teenage girls and women in their twenties.

1.2 Types of Fibers

As the historic examples showed, humankind has used a variety of fibers to create fabrics that provide thermal comfort for the conditions of their environment. Generally, we distinguish between natural, regenerated, and synthetic fibers.

1.2.1 Natural Fibers from Animals, Spiders[11], and Insects

Natural fibers from animals are qiviut, cashmere, angora, mohair, camel (Alpaca/Lama/Camel/Vicuna), and wool from sheep. Of course, wool properties differ among the species. The softest fiber, for instance, comes from the Angora rabbit; humans have harvested its hair for hundreds of years. Angora fibers have high heat retention and the best moisture-wicking properties of any natural fiber. Since the hairs are all medullated (i.e., hollow), angora fabrics have low weight and high insulating properties. However, due to the very low height of the fiber scales (as compared to other animal fibers), angora fibers have very poor spinnability. Therefore, producing fine-spun yarn is difficult. Low-scale heights also mean a high risk of fiber shedding due to the lack of fiber-to-fiber friction.

The finest fiber stems from the *Vicuna*, a small, wild camel family. Consequently, this fiber is rare and very expensive. In general, camel wool is soft, lustrous, lightweight, and provides good insulation. Cashmere stems from the fine, soft undercoat of the *Kashmir goat*. It is lightweight, extremely soft, very luxurious, has excellent drape, and is expensive. Wool is the fleece of sheep. Wool fabric is wrinkle-resistant, lightweight, durable, moisture-absorbent, and elastic, among other things.

Qiviut is the downy, soft under-wool of the *Arctic musk ox* that sheds naturally each spring. Qiviut is not scratchy. It is easy to handle because it does not shrink, no matter the water temperature. A qiviut garment provides eight times higher insulation and is much lighter than a wool garment. Qiviut items are hand-knitted and equally comfortable in warm and cold weather.

Silk is secreted by the *silkworm larva*. The finest quality stems from the unwound filament of the silkworm cocoon. Harvesting the filament requires killing the larvae in hot water. Damaged cocoons, broken or waste filaments provide low grades of silk. Silk fabric has a high natural luster, retains its shape, and drapes well. It is naturally hypoallergenic, breathable, and comfortable in both summer and winter.

---

[11] Spider fiber is rarely used.

1.2.2 Natural Fibers from Plants

The most commonly used natural plant fibers are from cotton, flax, and hemp. Cotton is a soft fiber that grows around the seeds. It makes up about 54% of the world's fiber production [10]. Cotton fabric naturally takes up and spreads sweat. However, it wrinkles and creases easily. Furthermore, cotton fabrics tend to shrink.

Bast fibers are strong, soft, woody fibers from the inner bark in the stems of plants like flax, jute, hemp, banana, and ramie. Hemp is the most durable and strongest of all natural textile fibers. It softens with each washing without fiber degradation. Moreover, it is resistant to mildew, mold, rot, and ultra-violet (UV) radiation. Because it takes up water easily, hemp is easy to dye even with natural dyes. Hemp is breathable and anti-microbial. However, hemp wrinkles easily and drapes poorly. Because of its ecological benefits (biodegradability), the fashion industry reintroduced hemp as a fiber source for fabrics. *Cannabis Sativa* provides the highest quality hemp, while *Sisal* and *Manila* hemp (aka *Abaca*) are of lower quality.

Obviously, the term *linen* originally referred to the woven fabric of any plant-based fibers, including hemp/flax blends[12]. However, in the 18th century, both flax and hemp lost their importance [8] due to, among other things, American laws that favored the growth and trade of cotton [11]. Linen is made from the cellulosic fibers of the stems of the flax (*Linum usitatissimum*) plant. It is much stronger and more lustrous than cotton. It feels cool to touch and is absorbent, but also wrinkles easily.

Pure natural (virgin) bamboo fiber stems directly from the bamboo stalk. Natural bamboo fibers are similar in performance to ramie, hemp, and flax. Furthermore, they are UV-and-wrinkle-resistant, antibacterial, highly hygroscopic, and yet exhibit breathability.

Today, banana-stem and pineapple-leaves fibers are rarely used because of the labor-intensive costs. However, fibers of pineapple leaves (*ananas comosus*) yield lightweight, breathable fabrics with marvelous cooling properties ideal for the Tropics. The fiber captures natural dyes very well.

Banana (*Musa paradisiaca*) is one of the oldest cultivated fruits in tropical and subtropical regions. The plant's pseudo-stem has three types of fibers. Because of its toughness, fabrics from fiber of the

---

12 Today the term linen often refers to cloth from cotton.

outermost layer are best for slacks, pants, dresses, and skirts. The middle layer serves to produce ropes, mats, and thick cloth. The innermost layer has the silkiest fibers. Therefore, fabrics from these fine fibers are great for T-shirts, blouses, cardigans, shawls, scarves, or undergarments [12]. Yarn from banana fiber is bio-degradable.

1.2.3 Regenerated Cellulose Fibers

The polymers of *regenerated fibers* stem from the cellulose of plants. They can be cellulose ester fibers, protein, or miscellaneous natural polymer fibers. Fabrics from regenerated fibers include acetate, bamboo, cellulose viscose rayon, Cupram-monium rayon, hydrolyzed acetate rayon, *Modal*, nitro-cellulose rayon, rayon, secondary acetate, *Tencel* aka *Lyocell*, and triacetate. Technically, these fibers are *semi-synthetics*.

Georges Audemars created the first artificial silk fibers in the mid-1850s. Louis-Marie Hilaire Bernigaud de Grange, Comte de Chardonne invented *viscose rayon* in 1883. Viscose rayon became a cheap alternative to silk due to its similar drape and texture. In 1892, Charles Fredick Cross and Edward John Bevan received the patent in Britain. Originally, rayon was produced from purified regenerated wood cellulose (cellulose acetate), which could include bamboo cellulose. The cellulose can also contain pulp-related agricultural products. The viscose process encompasses forcing the cellulose pulp through a so-called *spinneret* (a device with tiny holes) which creates pulpy filaments. These filaments then coagulate[13] in an acidic solution. Viscose rayon has low receptiveness to dyes.

The invention of *cellulose acetate* dates back to 1918. Its production involves adding glacial acetic acid, acetic anhydride, sulfuric acid, acetone, and water to the purified cellulose. The result is white yarn. After weaving, acetate fabrics have a very good drape, a soft, cool, and smooth feel, and low absorbency. In apparel, cellulose acetate or blends thereof are used for dresses, sportswear, shirts, ties, and underwear.

There exist mechanical and chemical ways to create *viscose-bamboo* fiber. The mechanical process is time-consuming and labor-intensive similar to that of flax. Furthermore, the polymerization process is very slow due to mild chemicals. The chemical route is similar to that of

---

[13] Coagulate means that particles suspended in a liquid precipitate out. Typically, this process forms a gel.

synthetics, except that the raw material stems from bamboo. Chemicals used at different steps in the process are a 20% solution of sodium hydroxide (NaOH), alkali, and carbon sulfide. Bamboo viscose appears like silk in sheen, drape, and sensation[14]. Bamboo fabrics are popular because they are bio-degradable and from sustainable sources.

Modal is a rayon made from 100% beech-tree pulp. *Lyocell* is a type of rayon from a wood-cellulose fiber developed by *American Enka* in 1972. Lyocell was originally trademarked as *Tencel* in 1982. Today, the term lyocell also refers to processes producing cellulose fibers. These processes involve dissolving the pulp, use of synthetic substances, and dry or wet spinning.

In *dry spinning*, the solvent evaporates after extrusion of the dissolved material through a spinneret. This process produces continuous filaments. *Wet spinning* involves extruding the spinning solution into a liquid-coagulating bath[15]. Here, the filaments form. They are then drawn and dried.

In the 1990s, a new process was developed for Tencel- and Modal-fiber production in response to the health-adverse impacts on workers. The new process includes the recovery and reuse of the amine oxide added to the wood pulp for polymerization [13]. Tencel and Modal fabrics are soft with a leather-like touch, have good wicking properties and very good breathability, and keep the skin cool and dry.

All fabrics from regenerated fibers are wrinkle-resistant and anti-static. Furthermore, they have higher uniformity, a brighter luster, a softer feel, and are more absorbent than fabrics from natural fibers. All fabrics from regenerated fibers are considered sustainable, and "eco-friendly."[16]

---

[14] Products made from viscose-bamboo dominate the market due to their lower price and easier care as compared to virgin bamboo. Unfortunately, due to labelling issues, many customers confuse garments produced from natural bamboo and regenerated viscose bamboo.

[15] A coagulation bath is applied in wet spinning of rayon or acrylic fibers after extrusion through a spinneret. This liquid bath hardens viscous polymer strands into solid fibers.

[16] Medical studies suggest that the carbon disulfide used in viscose production has health-adverse effects on occupational workers (e.g.,[12], [101]).

1.2.4 Synthetic Fibers

Synthetic fibers encompass acrylics, nylon, elastomeric fiber, polyesters (PES), polyolefins, polyurethane, and vinyl. Their polymers stem from petroleum.

In 1939, Wallace Carothers produced the first totally synthetic fiber, called *nylon*. Nylon consists of polymers with a protein-like structure that can, among others, build filaments suitable to produce yarns. In the fashion industry, nylon refers to tough, lightweight, elastic fabrics made from polymer.

Acrylic fibers are wool-like synthetic fibers. In 1944, *DuPont* produced the first acrylic fibers under the trade name *Orlon*. Shortly thereafter, *Monsanto*, *Hoechst*, and *Bayer* developed acrylic fibers under the trade names *Acrilan*, *Dolan*, and *Dalon*, respectively. The fiber-forming substance can be any long-chain synthetic polymer with at least 85% by weight of acrylonitrile units. Both dry and wet spinning (extrusion) can be applied alternatively.

Due to their versatile applicability, acrylic fibers have become one of the most commonly used synthetic fibers worldwide. Of all acrylic fibers produced, 75% go into apparel for cardigans, jackets, jogging suits, jumpers, knee-high stockings, socks, training suits, and waistcoats. In knitted garments, acrylic fibers are one of the most commonly used synthetic fibers because of their insulation, good shape-retaining, easy-care, softness of touch, quick-drying, and durability properties. Furthermore, acrylics are resistant to chemicals, moths, oil, and sunlight.

In the clothing industry, the most often used synthetic fiber is *polyester*. Despite polyester is a synthetic product, it may have plant origin (e.g., soy, milk). Synthetic fiber from milk protein was first introduced in Italy and in the United States in 1930. The main differences between polyester and acrylic are that the former is more breathable, and the latter is a better insulator.

1.2.5 Blended Fibers

Key fiber and fabric functional aspects encompass the suitability for the purpose, thermo-physiological comfort for the weather of the region for which the garment is designed, as well as the ease of production and care. Economic aspects are material and production costs and price value (e.g., *cost-per-wear*), which depend on the fabric's durability and response to wear and tear.

For many purposes, neither natural nor synthetic fibers are optimal. However, one can modify yarn properties by blending fibers. In other words, fabrics from blends can benefit from the properties of both fibers. Herein, the blending ratio can achieve different properties. Consequently, the fashion industry can address different requirements. For instance, wrinkle resistance and the ability to recover from deformation have made wool a favorite choice for apparel. Wrinkle-free single-jersey fabrics can be created by blending cotton with just 5% wool.

Blending acrylic with animal fibers increases fabric thickness and weight due to the increasing yarn hairiness with increasing animal fiber content. An acrylic/wool blend results in a wrinkle-resistant fabric with moisture absorption and good insulation. Often acrylic is blended with luxury animal fibers to obtain superior tactile properties despite spinning difficulties and production limitations for these blends.

Due to the low spinnability of Angora hair, for instance, angora is often blended with wool [14]. In these blends, the evenness of yarn decreases as the proportion of angora increased. Pure angora yarn has more irregularities than pure merino-wool yarn. Yarn with a 35% angora 65% merino wool blend ratio has better aesthetic appeal, appearance, and lower cost than pure angora. In the following, the mixing ratio of blends is given as X/Y fiber1/fiber2. Applying this convention to the aforementioned blend would read 35/65 Angora/Merino.

Blending fibers can improve the feel to touch or durability. For instance, hemp can feel harsh. Therefore, it is often blended with cotton, silk, wool, or polyester. Blending flax with polyester, for instance, especially enhances the mechanical properties of the fabric. Flax/polyester yarn becomes more regular (even) with increasing polyester content [15]. Blending yarns with *Lycra* or *spandex* improves fit, elasticity, and comfort.

1.3 Types of Fabric Structure

We distinguish various knit and weave patterns (Fig. 1.4). In knitting, the total number of loops in horizontal rows is called a *course*. *Wale* refers to the total number of vertical rows. *Course density* gives the number of visible loops per unit length counted along a wale. Analogously, *wale density* gives the number of visible loops per unit length counted along a course. *Stitch density* is the product of course density times wale density. Therefore, stitch density gives the change of courses and wales per unit area. The *yarn count* (number) gives the length of yarn in relation to the weight. Therefore, the yarn count affects fabric weight and thickness.

In *weft knitting*, one continuous thread runs crosswise in the fabric, making all of the loops in one course. Examples include jersey, double-knit, circular knitting, cable knit, and Fair Isle. Single-knit jersey, for instance, requires fine yarn on a knitting machine with just one row of needles using a *stocking stitch*. Stocking stitch, stockinette stitch, or plain knitting refers to applying alternatively one row of *knit* and one row of *purl*. The resulting fabric shows V-like and crescent shapes on the front and back, respectively.

A *pineapple knit* typically consists of rows 1 and 3 in knit, rows 2 and 4 in purl. In row 5, two loops knit, two loops knitted together using a knit stitch. In row 6, two loops purl and creation of a new loop in purl. This basic pattern is repeated. Due to knitting two loops together, the piece gets large lace-like holes (Fig. 1.4).

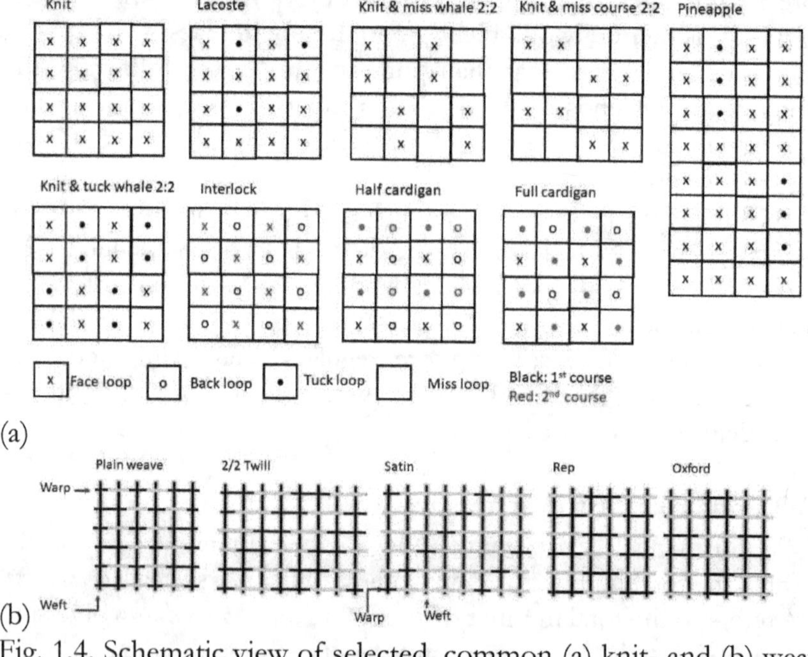

Fig. 1.4. Schematic view of selected, common (a) knit- and (b) weave pattern.

There also exist location-specific patterns. Knitters from Nightmute, Tununak, Newtok, and Toksook Bay (all remote coastal villages in Alaska), for instance, knit the *Nelson Island Diamond pattern*. This pattern has its origin in their traditional design of parka trims. The inspiration for the *Harpoon pattern* from Mekoryuk on Nunivak Island is a 1,200 years-old ivory harpoon head found on the island. These patterns are copyrighted for the use of the Oomingmak[17] members, and are not for sale.

Milano refers to a run-resistant *warp knitting* with a diagonal rib effect produced with several sets of yarns for each needle. In warp knitting, yarns run lengthwise in the fabric. Other examples of warp are *tricot* and *raschel*.

Double-knit or *bi-layer knit* structures like *interlock* (Fig. 1.4) are produced on two needle beds. Consequently, two layers are formed. Interlock-knit fabric is a variation of rib-knit. The two rows of needles cross over each other to construct the two rows of stitches. The rows become "interlocked" during the knitting process.

In weaves, we distinguish between 2-dimensional (2D) and 3-dimensional (3D) fabrics. 2D-woven fabric consists of two yarn sets: *warp* (0°) and *weft* (90°). Interlacing of the warp and weft yarns creates the fabric surface. Examples of 2D-woven fabrics are chiffon, crepe, denim, and satin[18]. In plain weave, weft and warp yarns are arranged 1/1, with the warp yarn alternating between being under or over a weft yarn at each crossover point (Fig. 1.4). In n-harness or $n/1$ satin or n-end satin, warp yarns run over n-1 weft yarns and under one weft yarn. Typically, n is equal to 4, 5, 8, or 12. This arrangement creates a smooth front and matte, dull reverse because there, the weft threads dominate. A satin weave is more elastic and more stretchable than a plain weave. Twill shows diagonal lines due to an offset in the warp yarns (Fig. 1.4). Garment examples are jeans and chinos. Twill is popular due to its durability and low staining.

Seersucker and *piqué* are examples of 3D fabrics. Piqué has two warps and two filling yarns, one fine, and one heavy. This double cloth is a medium- to heavy-weight fabric with raised cords in warp direction. Seersucker, aka *crimp effects*, refers to cotton fabrics with

---

[17] Oomingmak is a Native Alaskans cooperative founded in 1969. Their members hand-knit qiviut scarves, stoles, smokerings (aka *nachaqs*), hats, and tunics.

[18] Satin weave made of cotton is called *sateen*.

puckered and flat strips in warp direction. Its major advantage is that it needs no ironing.

1.4 Why We Have to Dress — The Laws of Thermodynamics

Obviously, some people only dress to not die from hypothermia in winter, and to not get into trouble with the police in summer. Actually, besides style and self-expression, there are thermodynamic, energetic, and health reasons like insulation, cooling, sun exposure, etc. Accept the fundamental laws of thermodynamics as *axioms* that define a group of physical quantities. Axioms are not further explainable, but matters of facts. The *zeroth law of thermodynamics* defines *thermal equilibrium* and *temperature*. When two bodies both in their own state of internal thermodynamic equilibrium are in thermal equilibrium with each other, they have the same temperature. So, to speak, the zeroth law of thermodynamics directs us on how to measure temperature — by creating a thermal equilibrium. Recall this simple everyday example. You measure fever with a thermometer by putting it under the arm and waiting a while. The waiting time serves to establish the thermodynamic equilibrium.

The *first law of thermodynamics* postulates that when energy (in the form of work, heat, radiation, or matter like food) flows into/out of a system, the internal energy of the system changes in accordance with the law of *energy conservation*[19]. Let me give you an easy example; when you eat more food than your body burns for your activities, your body stores the excess energy in the form of matter, aka body fat.

The *second law of thermodynamics* states that heat does not spontaneously flow from a colder body to a warmer body. In other words, it tells us the direction in which *net energy* flows. A warmer body (e.g., the Sun) has a higher energy than a colder body (e.g., the Earth). While they both radiate according to their temperature, the warmer body receives less energy from the colder body than the colder body receives from the warmer. Consequently, the *net energy* (i.e., the difference between received and emitted energy) flows towards the cold body.

---

[19] The law of energy conservation says that energy is neither created nor destroyed. In other words, when you use energy, energy just changes from one form of energy into another form of energy. This means it won't disappear.

In simple words, humans' need for weather-appropriate apparel is a consequence of the second law of thermodynamics. When namely the environment has a lower temperature than the body temperature, a person loses heat; when the opposite is true, the human body gains heat. Both can cause death in the extreme case. Therefore, we have to dress appropriately to avoid undercooling or overheating.

The *third law of thermodynamics* defines the absolute zero point of the absolute temperature (in Kelvin) as the system's *entropy* approaching zero. Herein, entropy is the system's availability of *thermal energy* that can be converted into *mechanical work*. In simple words, the third law of thermodynamics gives us a **fixed reference point** that allows us to measure the absolute entropy of any substance at any temperature. In this book, we only use this physical concept to define the *Kelvin-temperature scale*. This temperature scale is typically applied in science and engineering.

1.4.1 Relation between the Celsius-, Fahrenheit-, and Kelvin Scales

Whenever you want to measure something, you have to establish a scale. To do so, you choose two reference points that you can use for calibration. A reference point must fulfill the criterion that you will get the same value when you repeat the observation under the same conditions. Then you perform all your measurements with this scale, and you can compare your observations.

In 1724, Daniel Gabriel Fahrenheit, a physicist of Dutch–German–Polish heritage, developed a temperature scale and named the units after himself. Several stories exist about how he established his scale. The most logical one is that he created a solution of brine consisting of equal fractions of ice, water, and ammonium chloride and took this solution's temperature as the lower reference (0F). Obviously, the brine solution was the lowest temperature he knew. Some stories refer to the temperature of the human blood as the upper-level set at 100F. Actually, later the *Fahrenheit scale* was revised because the normal body temperature is actually 98.6F (37°C). Another story takes the melting point of ice (32F), and boiling point of water (212F) at normal sea level pressure (29.92 inches of Mercury=1013.25 mbar=1013.25 hPa=1 atm) as reference. However, setting these points to 32F and 212F, respectively, sounds not straightforward.

In 1742, the Swedish astronomer Anders Celsius created the *Celsius scale*. He defined the freezing and boiling points of water at normal sea level pressure as 0°C and 100°C, respectively.

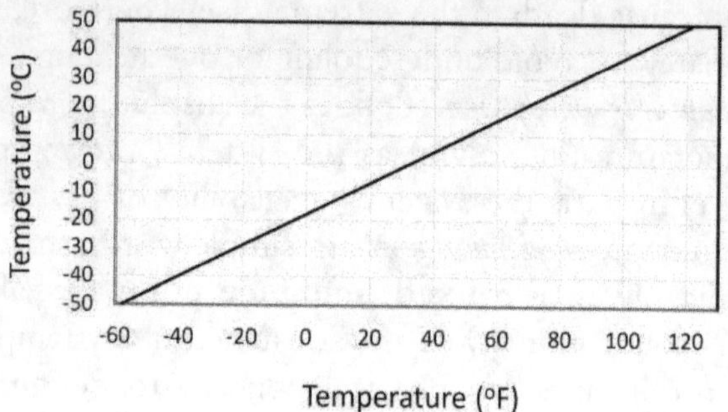

Fig. 1.5. Comparison of the Fahrenheit- and Celsius scales temperatures.

You can use the following formulas to convert temperatures from the Fahrenheit scale into the Celsius scale and vice versa.

F to °C: $T_C = (T_F - 32) \cdot 5/9\,°C$

°C to F: $T_F = (9/5 \cdot T_C + 32)\,F = (1.8 \cdot T_C + 32)\,F$

Here T_F and T_C are the temperatures in Fahrenheit and Celsius, respectively. The only point at which the values on the Fahrenheit and Celsius scales are the same is at -40 (Fig. 1.5). Adding 273.15 to a temperature given on the Celsius scale converts it to the temperature (T) on the *Kelvin scale* (K).

°C to K: $T = T_c + 273.15$

1.5 Why to Understand the Relation between Fashion, Fabric, Weather, and Comfort

Can we learn from the ancestors and adapt the underlying energetic concepts of clothing in our lifestyle in a way that we still look stylish when we try to protect ourselves from the cold/heat and sun, and stay healthy? The answer is definitely **yes, we can**. In other words, when you live in a cold climate, you do not need to dress like scientists in

Antarctica. Typically, the national funding agencies of all countries performing research in Antarctica require their scientists to wear heavily insulated clothing that includes thermal underwear, wind-proof leggings, heavy socks, an insulated vest, insulated parka, thermal boots or mukluks, cap, helmet, neck gaiter, various gloves, and goggles (Fig. 1.6). Such outfits are admittedly not stylish. And a polar bear looks like being on a strict diet compared to a scientist in their mandatory outfit. Probably, that is why none of my colleagues, who had been in Antarctica or on an Arctic research cruise, volunteered to give me a photo in full gear for this book.

Fig. 1.6. "US Ambassador in Antarctica 2010 - Day 2" by US Embassy New Zealand is marked with public domain mark 1.0.

However, dressing stylishly for thermo-physiological comfort requires understanding the energetics of the various fabrics in conjunction with the weather conditions and our metabolism. Let's discuss the attributes of clothing fabrics beyond sootiness vs. itchiness, softness vs. harshness, smoothness vs. roughness, warmness vs. coolness, and expensive vs. cheap. After reading this book, you can identify sun-safe fabrics and define thermo-physiological comfort in the framework of the system *environment-weather-clothing-body*. This knowledge helps you in buying the right clothes for your needs, dressing for the weather, and creating layers for thermo-physiological comfort. Fashion designers can create garments that people not only look great in but also feel comfortable in.

2 REQUIRMENTS AND DESIGN OF SUN-SAFE FASHION

2.1 The Sun, and Its Radiation

The Sun emits radiation similar to what would be expected from a 5778 K (5505°C, 9941F) *blackbody*[20]. This value is roughly the Sun's "surface temperature."[21] The *solar radiation* (aka *insolation* or *shortwave downward radiation*) transfers heat by electromagnetic waves. This energy from the Sun travels through space to the Earth. At the top of the Earth's atmosphere (TOA), the strength of the Sun's radiation differs by latitude, day (Fig. 2.1), time of the day, and hemisphere. As the diagram illustrates, at the TOA, daily mean solar radiation is lowest in the respective winter of each hemisphere. In a hemisphere's summer, it is the largest over its Polar region because this region receives solar radiation 24/7 (aka *white nights*). Consequently, summed up over the day, this Polar region experiences more solar radiation at the TOA on a summer day than the TOA of the Tropics on that same day. In winter, the days remain dark at latitudes above the Arctic (Antarctic) circle for an appreciable number of days, depending on the latitude. Consequently, the TOA receives no solar radiation in these areas (Fig. 2.1).

Figure 2.1 also reveals that the southern hemisphere receives more solar radiation than the northern hemisphere. The reason is the variation in the Earth-Sun distance as the Earth revolves around the Sun. The Earth's angle to the ecliptic is the reason for the seasons [16].

Once the solar radiation reaches the TOA, it has to travel through the atmosphere to reach the Earth's or your skin's surface. The distance the sunrays have to travel through the atmosphere depends on the altitude, time of the day, day of the year, latitude, the distance Earth-Sun, the Earth's ecliptic, and eccentricity, which changes at very long time scales [17]. The solar zenith angle gives the angle between the Sun at local noon (when solar radiation is the most powerful), and

---

[20] A blackbody is a hypothetical body that absorbs all radiant energy falling on it. Because a blackbody absorbs all incident visible light rather than reflecting it, its surface appears black.

[21] Actually, it's hard to define a surface in the sense of the Earth's surface for the Sun. Therefore, you find slightly differing values for the "Sun surface temperature" in the literature.

the center of the Sun. The solar elevation gives the angle of the Sun above the horizon. They both refer to the Sun's position in the sky. In simple words, the higher the Sun is in the sky, the shorter the distance the solar radiation travels through the atmosphere, and the more solar radiation can reach you. Consequently, at the same latitude, you are more exposed to the Sun at local noon when you are hiking on a mountain than at sea-level height or than early in the morning or late in the evening. Did you know that when you fly at 5 km height, there is already about 50% of the atmosphere's mass below you?

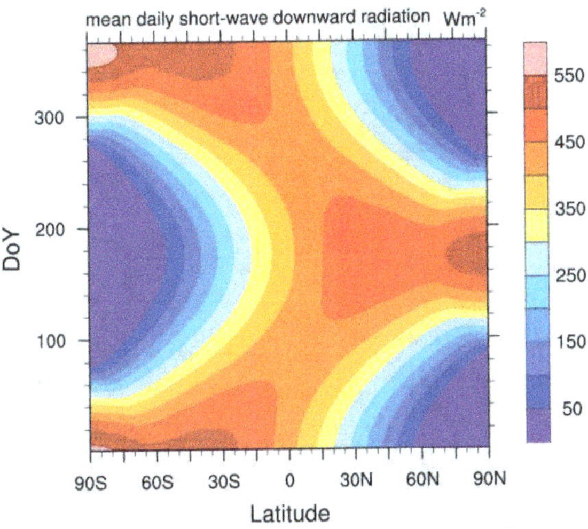

Fig. 2.1. Daily mean incoming solar radiation (aka *shortwave downward radiation*) at the top of the atmosphere. DoY is the day of the year. Counting starts January 1. From [18].

The Sun's *irradiance* is the radiation that is incident on a surface. It spans wavelengths from 100 nm ($3.9 \cdot 10^{-6}$ in) to 1,000,000 nm (1 mm ~0.04 in; Fig. 2.2). Commonly, we distinguish between the three ultraviolet (UV) – UVC, UVB, UVA –, visible, and infrared (IR) ranges. These ranges encompass wavelengths of 100-280 nm, 280-315 nm, 315-400 nm, 400-700 nm, and 700 nm to 1 mm, respectively. The visible spectrum is divided into seven separate bands that represent the colors red (650-800 nm), orange (590-640 nm), yellow (550-580 nm), green (490-530 nm), blue (460-480 nm), indigo (440-450 nm), and violet (390-430 nm). Technically, colors are how our brain interprets electromagnetic radiation of wavelengths located in the visible spectrum. See Ch. 6.

You have made yourself the experience that clouds dim down the sunlight reaching the ground. The thicker the clouds are, and/or the more the sky is cloud-covered, the larger is the effect. Cloudiness and its degree, but also absorbing gases, aerosols, dust, pollen, and air pollution, reduce the incoming solar radiation at the Earth's surface [16].

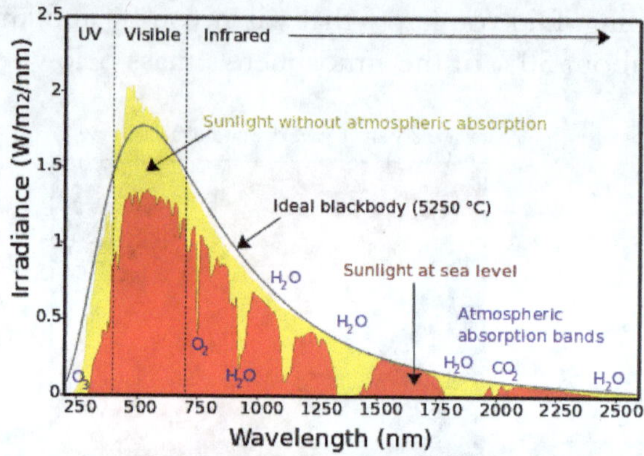

Fig. 2.2. Schematic view of the solar spectrum on Earth. Irradiance is the radiant flux received by a surface per unit area, i.e., the energy per unit time (e.g., second) that is radiated from the Sun over the optical wavelengths. The chemical formulas indicate the atmospheric gases that absorb solar radiation at that specific wavelength. Namely, ozone (O_3), oxygen (O_2), water vapor (H_2O), and carbon dioxide (CO_2). From: Nick84. Licensed CC BY-SA 3.0.

The black line in Figure 2.2 corresponds to the theoretical values of a perfect blackbody at the various wavelengths as obtained by *Planck's law*[22]. The yellow-shaded area is the sunlight that arrives at the TOA without absorption. Because various gases, particles, ice crystals, and cloud droplets absorb at different wavelengths, the spectrum of solar radiation reaching the Earth's surface differs from the theoretical spectrum predicted by Planck's law. The red-shaded area gives the solar radiation at the various wavelengths that arrive at a surface at sea-level height (e.g., on your skin). The difference between the area under the theoretical black line and the red-shaded area is what the atmosphere absorbs.

---

[22] Planck's Law gives the spectral radiance of electromagnetic radiation at all wavelengths from a blackbody at temperature T.

Now take a look at Figure 2.2 in the UV region. UVA light is not absorbed. In the wavelength range of 200 nm to 315 nm, the ozone layer absorbs, on average, about 98% of the UVB radiation. Absorption is slightly lower (higher) when there is less (more) ozone. Freon gases (i.e., gases used in spray bottles), among other gases, destroy ozone in the ozone layer [16].

Ozone is not the only gas that absorbs solar radiation in the UV range. Water vapor, carbon dioxide, as well as some aerosols can absorb UV light at different heights. Therefore, their vertical profiles of concentration and their horizontal distribution can influence your sun exposure as well. This means that you are more in "danger" of sunburn in the pristine air of a National Park or Wilderness Area than in a mega city.

Radiation is associated with energy. According to *Planck's equation*, the energy, E, reads

$$E = h\nu$$

Where $h = 6.626 \cdot 10^{-34}$ J/s is *Planck's constant*, and $\nu = c/\lambda$ is the *frequency* of the electromagnetic wave, λ, and $c = 3 \cdot 10^{8}$ m/s is the *speed of light* (in vacuum). Consequently, we obtain for the energy

$$E = h(c/\lambda).$$

This result means that the smaller the wavelength is, the higher is its energy.

Due to its high energy level, UV radiation can change the cell structure. Therefore, it is the most harmful to living species, followed by UVB and then UVA. Indeed, the so-called *shortwave ultraviolet, UVC,* or *germinal* (hard) UV light serves to sterilize surfaces and medical instruments. Fortunately, the ozone layer absorbs UVC (Fig. 2.2).

Because the wavelengths of UVB and UVA are longer than those of UVC (Fig. 2.2), they can reach the unprotected human skin. Here, UVA- and UVB rays can penetrate the *dermis* and *epidermis*, respectively (Fig. 2.3a). Because the epidermis is the top skin layer, UVB radiation is more effective in causing sunburn and *erythema*[23] than UVA radiation. However, all UV wavelengths damage collagen, which accelerates the

---

[23] Erythema refers to the reddening of skin.

aging of the skin. Both UVA and UVB radiation also break up vitamin A in the skin. Furthermore, UVB damages the DNA (Fig. 2.4), which can lead to skin cancer (e.g., [5]). Therefore, we need clothing that protects our skin from harmful UV radiation.

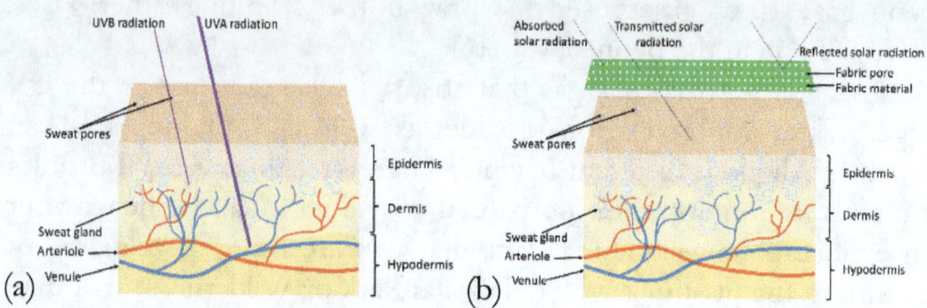

Fig. 2.3. Schematic view of (a) UVB and UVA radiation penetrating the human skin, and (b) the fate of solar radiation when reaching a fabric. Note that at the fabric-atmosphere interface, each photon can have only one of the three fates – being absorbed, transmitted, or reflected. A photon is the massless particle that, according to the quantum theory of radiation, carries the smallest discrete amount of electromagnetic energy, hν.

Radiation in the visible and infrared range can penetrate all three skin layers, i.e., also the lowermost skin layer called the *subcutaneous layer* or *hypodermis*. Because of their much lower energy, the wavelengths in the visible and IR range only heat the skin cells.

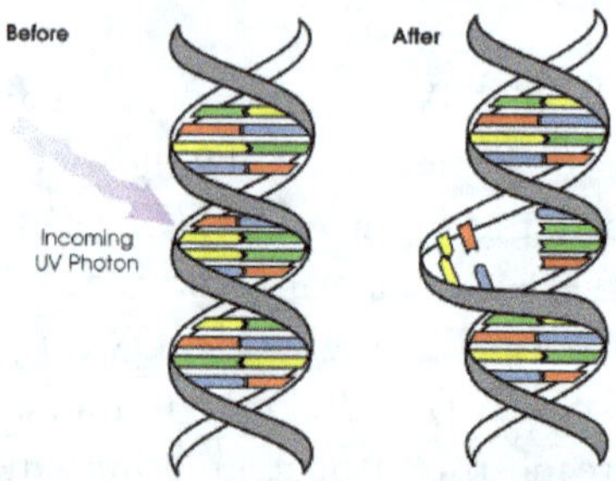

Fig. 2.4. Schematic view of DNA damage by UV light. Attribution: Derivative work: Mouagip (talk) DNA_UV_mutation.gif: NASA/David Herring. Public domain.

2.1.1 Albedo, Reflection, and Diffuse Radiation or Why We Get Sunburn in the Shade

Albedo refers to the ratio of the reflected sunlight of a given wavelength to the sunlight received at that wavelength. Typically, it is expressed as a fraction between 0 and 1. The reflection increases with increasing value of the albedo. Obviously, the albedo of a material can be different at different wavelengths. In the visible range, for instance, clouds, snow, ice, and white surfaces have a high albedo, but they are dark in the infrared [16]. Furthermore, in the visible range, albedo decreases when the material becomes wet. Just recall your experience that wet sand is darker than dry sand.

Cloud layers at different heights can reflect sunlight between each other in all directions. Furthermore, molecules and particles in the atmosphere, walls of buildings, leaves, the ground, etc., reflect sunrays (Fig. 2.5). Reflected rays can bounce back and forth between surfaces of various sizes (*multiple scattering*). In a small street, for instance, the houses along the street provide shadow. However, the rays reach walls, windows, or roofs at higher levels on one side of the street. Some rays heat the surface they hit, i.e., they get absorbed. Some rays bounce back, hitting the wall on the other side of the street. There they may again be reflected back, and so on. This means you are exposed to very diffuse sunlight, even in the shade (Fig. 2.5).

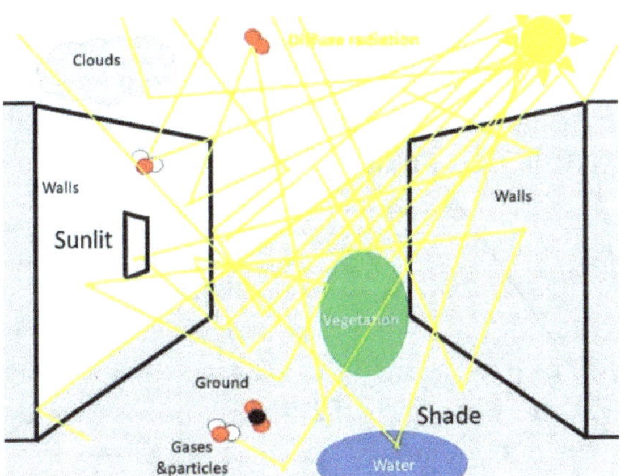

Fig. 2.5. Schematic view of diffuse radiation in an urban environment. Yellow lines are shortwave radiation, gray areas indicate shade.

Therefore, any shade-providing obstacles like an umbrella or trees protect you only from *direct sunlight*, i.e., the radiation that comes directly from the Sun. However, they fail to protect you from the reflected rays, the so-called *diffuse radiation* (Fig. 2.5). Unfortunately, this diffuse radiation also encompasses harmful UV rays that cause sunburn.

Now, what does the existence of diffuse radiation mean for you? Sunburns can happen at all times of the year, especially in environments with highly reflective surfaces like snow, water, white walls, mirrored windows, or the glass windows of a street mall. Due to reflective surfaces in the environment and in the atmosphere (clouds, molecules, particles), you not only receive the sunrays directly coming from the Sun, but also sunrays from diffuse radiation. Therefore, following the general rule of thumb, "Watch your shadow. Short shadow, seek shade." only reduces the direct sun exposure. Diffuse UV radiation is the reason why being in the shade fails to rescue you from sunburn.

In a nutshell: There are at least ten factors that affect your exposure to UV radiation namely, the time spent in the Sun, the solar zenith or elevation angle, your altitude, latitude, the day and time of the year, the reflection from surfaces, the ozone layer, atmospheric trace gases, water vapor, atmospheric particles (e.g., aerosols, pollen, dust, cloud-ice crystals, cloud-water droplets), using or not using sunscreen, and **what you wear**. It is a myth that you are safe from sunrays in the shadow and/or behind glass. Indeed, you can get a terrible sunburn in your car or in a glider plane, especially when you are up high. Shade only mitigates the *heat stress*, but fails to avoid sunburn.

2.2 UV Radiation and Fabric Porosity

There is an old myth that UV radiation cannot pass through fabrics. Many of us have learned the hard way that this belief is wrong. Fabrics, namely, are porous media. Fabric porosity is defined as the ratio of open space to the total volume of the fabric, V_{fabric}. The fabric density-based porosity, η_{fabric} reads

$$\eta_{fabric} = (V_{fabric} - m/\rho_{fiber})/V_{fabric}$$

Where $V_{fabric}=A \cdot h$ is the product of the fabric area, A, and fabric thickness, h (Fig. 2.6). Furthermore, m and ρ_{fiber} are the fabric weight (aka *mass*) per area and fiber density, respectively. Obviously, fabric porosity varies with fabric structure, material, thickness, treatment, weight per unit area, yarn twist, and the tightness of weave/knit. This means when fiber density, dimension, and shape differ, two fabrics with similar woven structures and geometry can have distinctly different porosity.

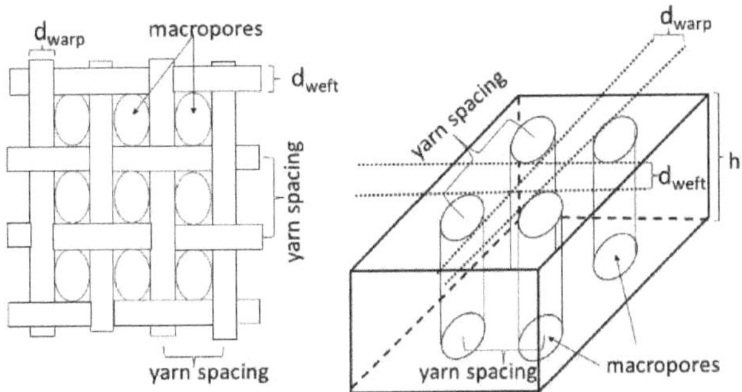

Fig. 2.6. Schematic 2D- and 3D-view of fabric porosity. Typically, yarn spacing differs in weft and warp direction. Therefore, the macropores are shown as ovals. Furthermore, d is the yarn thickness which can differ in weft and wrap direction.

We can calculate fabric bulk density, ρ_{fabric}, by multiplying the fabric thickness (in meters) with the fabric weight per square meter. Consequently, high porosity means many small air pores. Furthermore, a low fabric bulk density means more air-filled pores exist between the fibers of the fabric, and theoretically, more possibilities exist for sunrays to travel through.

We also know from experience that everything that is smaller than an opening can travel through it. For instance, a mosquito can fly through a mesh fence, but an owl can't. This example shows us that the ratio of the traveling object to the size of the openings plays a role. This means radiation of wavelengths smaller than the pores can travel through the fabric (Fig. 2.3b). Scientists call this fraction of incident electromagnetic power that travels through the fabric, *transmittance*. Due to the transmittance, we have to answer which wavelengths can penetrate through which fabrics.

2.2.1 Rating Sun-Protective Clothing

Theoretically, fabric can transmit, absorb or reflect radiation (Fig. 2.3b). Some fibers tend to absorb a fraction of the UV radiation, and convert it into heat. Another fraction is reflected in all directions. In case of non-sun-safe fabrics, the fabric transmits a substantial fraction of the UV radiation to your skin. In simple words, this fraction of the UV radiation penetrates directly through interstices between the fibers and yarns of the fabrics. In turn, your risk of sunburn, and your risk for skin cancer, may increase [19].

To rate the UV protection of sunscreens and fabrics, the international scientific community has established the *Ultraviolet Protection Factor (UPF)*. The UPF value is 100 divided by porosity. The UPF measures how much of the UV radiation falling on a fabric travels through the fabric (Fig. 2.7). A UPF50 value, for instance, means that only $1/50^{th}$ of the UV radiation falling on the fabric passes through it. Or in simple words, the fabric blocks $49/50^{th}$, which is 98% of the UV radiation.

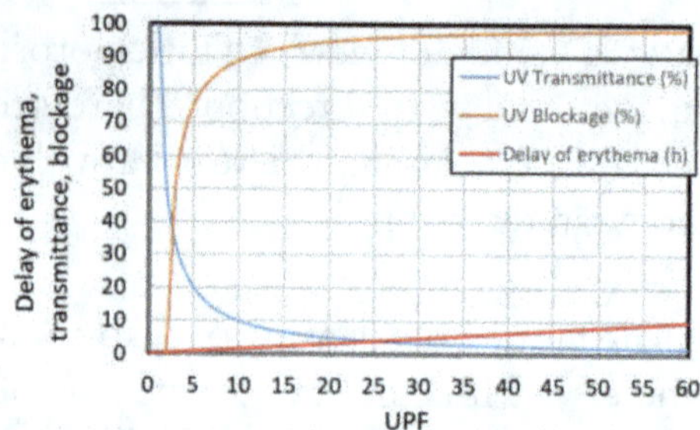

Fig. 2.7. Percentage of UV transmitted through and blocked by fabrics of various UPF values, and hours of delay of reddening at the various UPF values.

So-called *lab-sphere's UV transmittance analyzers* measure the UV radiation blocked and transmitted by fabrics. Then the UPF is calculated as the ratio of the erythemally-weighted UV-radiation flux of radiant energy per unit area (aka irradiance) at the detector without and with fabric as

$$UPF = \Sigma\, E_\lambda \cdot S_\lambda \cdot T_\lambda \cdot \Delta\lambda / (\Sigma\, E_\lambda \cdot S_\lambda \cdot T_\lambda)$$

Where E_λ, S_λ, and T_λ are the relative *erythemal spectral effectiveness*, solar spectral irradiance, and average spectral transmittance of the fabric measured at wavelength λ in the wavelength range, $\Delta\lambda$. Furthermore, $\sum$ is the summation sign. Here the sign stands for the sum of $E_\lambda \cdot S_\lambda \cdot T_\lambda \cdot \Delta\lambda$ over all wavelengths λ. In simple words, $\sum$ is an abbreviated way for writing $E_{\lambda 1} \cdot S_{\lambda 1} \cdot T_{\lambda 1} \cdot \Delta\lambda_1 + E_{\lambda 2} \cdot S_{\lambda 2} \cdot T_{\lambda 2} \cdot \Delta\lambda_2 + \ldots + E_{\lambda n} \cdot S_{\lambda n} \cdot T_{\lambda n} \cdot \Delta\lambda_n$ where 1, 2, ... n denote the wavelength ranges considered.

Science suggests UPF ratings of 40+ or 50+ as being "totally safe" or "sun-blocking," respectively. Or explained in a simple way: If a normal fair-skin person exposed continuously to UV radiation starts reddening after 10 minutes without a sun shield, using a UPF15 sun shield would delay the onset of reddening to 150 minutes. Consequently, the person could stay 15 times longer in the Sun before reddening would occur. UPF values ranging from 15-24, 25-39, and greater than 40 are considered good, very good, and excellent, respectively [20].

Did you know that one-third of commercial summer clothing provides a UPF of less than 15? How do you recognize such clothing? Well, hold the fabric against the Sun. When you see the Sun through the fabric, UV rays can penetrate through the fabric. Why? Because UV radiation has shorter wavelengths than the visible radiation your eyes can see (Fig. 2.2). Why are summer clothes prone to transmit UV rays? They are light, and, typically, fabric weight dominates the effects of fabric structures on UPF.

2.2.1.1 Cover Factor

Stich density of knitted fabrics and weave density (fabric structure) influence the porosity and coverage of the skin (Fig. 2.6). We can define the *cover factor*, CF, as the fraction of surface covered by yarns. It corresponds to the quotient of area covered by yarns and the total area of web. Consequently, the cover factor varies between 0 and 1. Multiplication with 100 gives the cover factor in percent.

We can now determine the percentage of UV transmission through a fabric by 100% minus the cover factor. This means UPF=100/(100-CF). Rearranging for CF shows us that to obtain a minimum UPF of 15, for instance, the cover factor must exceed 93%. Because CF is in the denominator, a small increase in CF substantially improves the UPF. Typically, woven fabrics have a higher cover factor than knits.

Due to the irregularities in the fabric construction, UPF of terrycloth varies widely.

Obviously, the cover factor decreases linearly with increasing porosity for the same fabric structure and material (Fig. 2.8a). On the contrary, UPF nonlinearly increases with porosity (Fig. 2.8b) because transmittance through small pores requires shorter wavelengths.

From the concept of cover factor, it becomes obvious that stretch might alter a fabric's effectiveness of UV protection. Consequently, sun-protective clothing should fall loose on the body. Wetness alters the sun-protection too, because radiation can travel through water[24]. In plain English, stretched or wet fabrics may fail to cover your skin completely or to absorb UV radiation.

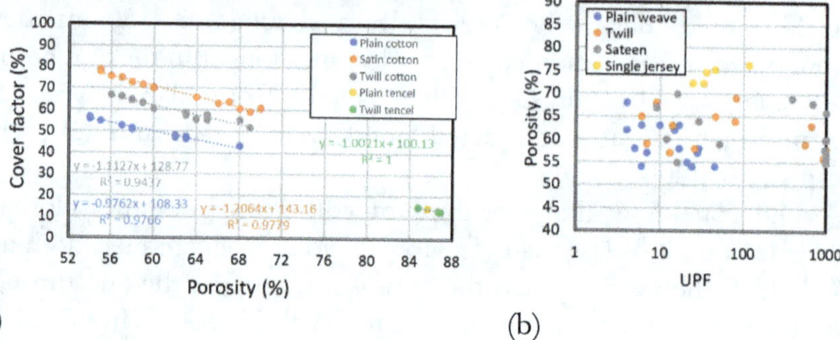

(a)   (b)

Fig. 2.8. Relationship between (a) cover factor and porosity, and (b) UPF and porosity for various weaves. In (a), the equations give the relation between porosity and cover factor, and R^2 multiplied by 100 is the percentage of correlation. Data from [21], [22], [23].

2.2.1.2 Impact of Fabric Structure and Porosity on UPF

In knits, machine gauge, yarn parameters like yarn count, and knit structure affect stitch density. Typically, pore size decreases with increasing stitch density, which depends on the knit pattern (Fig. 2.9). Obviously, fabric density increases with decreasing fabric porosity. The rate of increase differs among fabric materials and blends.

In fabrics, pores are not straight paths like the pores of a mosquito screen. According to the literature, the path lengths of fabric pores are about twice the thickness of the fabric for summer clothes. Therefore, textile engineers say that the fabric has a *path factor* of 2. The average

---

[24] You may have made the experience that you can still see under water when you snorkel.

distance between your skin and your summer clothes has a path factor of about 1.5.

Because porosity, among other things, depends on stitch or weave density (Fig. 2.9), producing fabrics with the maximum number of yarns in warp and weft directions results in high UPFs. For similar fabric construction, UPF increases with fabric thickness and density, and depends on porosity (UPF=100/porosity) [49]. Mean UPF increases with increasing stitch density in single-knit fabrics except for plain and pineapple patterns (Fig. 2.10).

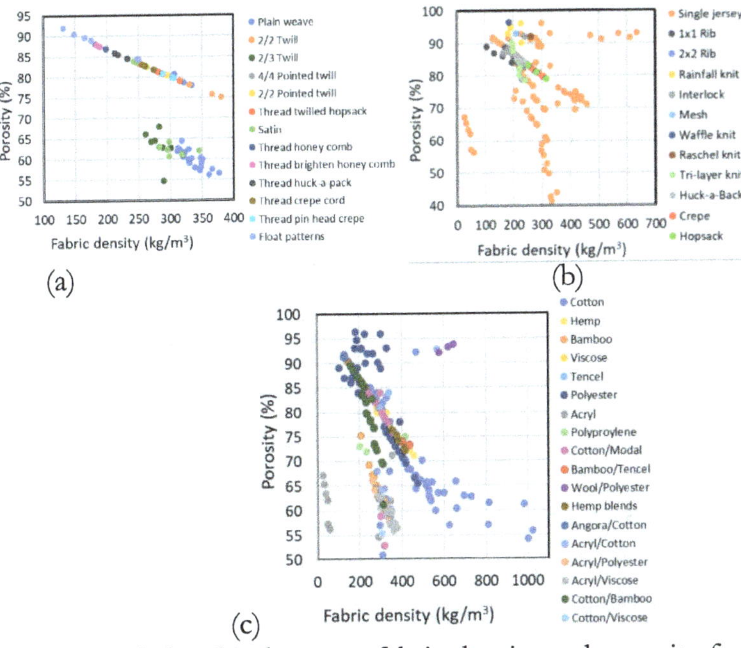

Fig. 2.9. Relationship between fabric density and porosity for various (a) weaves, (b) knits, and (c) fabric materials and blends. Data from [14], [21], [23], [24], [25], [26], [27], [28], [29], [30], [31], [32], [33], [34], [35], [36], [37], [38], [39], [40], [41], [42], [43], [44], [45], [46], [47], [48].

Fabric structure strongly affects porosity and fabric density. As a result, UPF depends on the tightness and type of the weave or knit (Fig. 2.10). To better understand how fabric structure affects UPF let's compare, for example, a *knit-and-miss pattern* with a *knit-and-tuck pattern* (Fig. 1.4). In a knit-and-miss pattern, the miss loops pull the knitted loops close together thereby increasing the stitch density. On the left side, the missed loops contribute to hindering UV radiation from penetrating through the knit. On the contrary, to create a tuck, a stitch from a few rows below is knitted together with the current stitch on the needle. As a result, the tuck loop is stretched. Due to these differences, a miss-and-knit pattern yields a higher UPF value than a knit-and-tuck pattern (Fig. 2.10).

Single-knit structures show a positive correlation between fabric weight and UPF. Therefore, we can say UPF increases with the increasing weight of the single knit fabric. In general, the relationship between fabric weight and UPF is higher for knit-tuck-miss structures than plain and pineapple structures. While the strength of the UPF-fabric-weight relationship varies among the various single-knit patterns, it differs strongly among double-knit structures, and becomes even slightly negative for *half-Milano*. However, the UPF decrease with increasing weight of half-Milano knits is too weak to explain the altered UPF by the change in fabric weight [50].

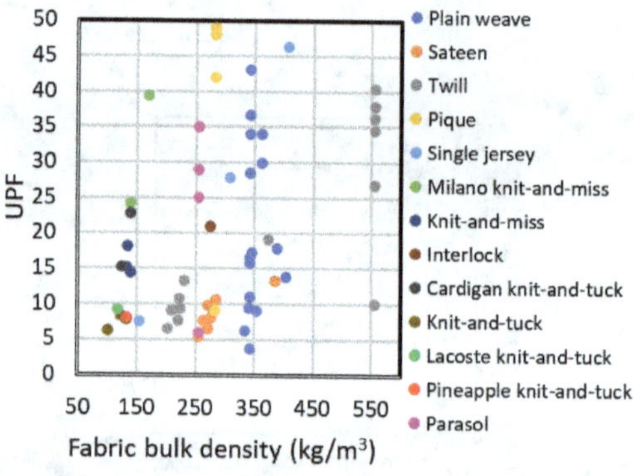

Fig. 2.10. UPF as a function of bulk density for selected cotton weaves and knits. Milano, interlock, cardigan, and 1×1 rib are double-knit structures. Data from [23], [50], [51], [52], [53].

2.2.1.3 Impact of Fiber Type and Yarn Configuration

Besides fabric structure, the yarn material also affects UPF. For instance, 10.52–12.70% and 9.03–11.47% of the UVA and UVB radiation can penetrate through thick rib hemp and linen fabrics, respectively. A knitted blend of synthetic fibers with Lycra best protects against solar radiation [49].

Yarn configuration is an important factor for UV protection of multifilament woven polyethylene terephthalate (PET) fabrics. Flat yarn without twist provides the most compact structure from the aspect of UV protection because each filament in flat yarn with zero twists is free to tie into any free space. Consequently, compactness is maximized. Intermingled yarn is second with respect to the compactness of the fabric structure, followed by textured multifilament PET yarns. The reason is the basic geometry of these yarns. A wavy filament structure permits high UV transmission through the yarn itself. The very compact structure of twisted yarns leaves maximum porosity when used for warp and weft fabrics [54]. Consequently, yarn configuration remarkably affects the UV protection characteristics of woven fabrics due to the effectiveness of *filament-in-fiber* packing of multifilament PET yarns. However, the compactness can affect the wear comfort negatively (Ch. 3). Therefore, designers have to choose the appropriate combination of warp and weft threads to produce maximum UV protection under consideration of the desired wear-comfort properties. This choice may differ depending on the purpose of the garment (e.g., sportswear, resorts wear, swimwear), and climate region.

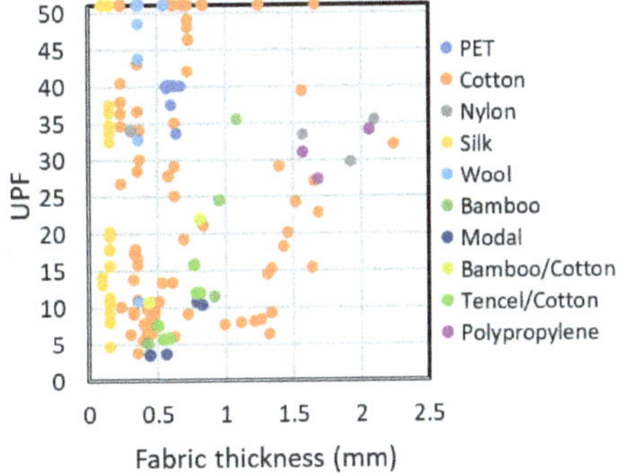

Fig. 2.11. UPF vs. fabric thickness for fabrics from various fibers. Note that for the same fiber type, the fabric treatment, structure, and/or tightness may differ. Values plotted at UPF=51 have UPF values greater than 50. Data from [20], [23], [50], [51], [53], [54], [55], [56], [57], [58], [59].

Obviously, at the same fabric thickness, the treatment and fabric structure strongly affect the UPF (Fig. 2.10). Typically, the importance of UV protection decreases in the order cover factor, fiber type, and fabric thickness. Due to its anti-ultraviolet nature, bamboo fiber is very suitable for summer clothing.

2.2.2 Relation between Fabric Construction and Reflectance

As aforementioned, across the electromagnetic spectrum, different materials have their unique reflectance and emission properties that depend on the wavelength. Herein reflectance gives the percentage of the total radiant flux incident upon a surface that is reflected. This percentage differs according to the wavelength distribution of the incident radiation.

Consequently, for protection from UV radiation, fabrics have to have high reflectance at wavelengths in the UV range. In general, the reflectance of fabrics depends on the cover factor, yarn-fiber fineness, yarn count, yarn density, and fabric construction (weave, knit). Polyester fabrics woven in plain and satin weave (Fig. 1.4) with the same yarn and yarn count have the same fabric-cover factors. However, their light reflectance differs from each other due to the changes in cover of the different weave types [60].

2.3 Enhancement of UPF with Dyes

We discussed already that shrinking of fabrics in hot water decreases the pore size (Ch. 1). In the production of *lodden* or *molten*, cooking of the fabrics served to achieve better protection from wind and rain. A side effect of the reduced space between interstices is an increase in UPF. In simple words, a mishap of shrinking your sweater in the laundry benefits your sun protection – pre-assumed the sweater remains loose enough.

Dyeing of fabrics affects porosity by shrinking and deposition of pigments into or on the yarn. Consequently, dye alters the cover factor (Fig. 2.11). A comparison of Figures 2.8 and 2.11 reveals a higher impact of fabric structure on UPF than of color.

In general, undyed – so-called *gray* or *greige* – fabrics protect less than fabrics dyed with synthetic or natural dyes (Fig. 2.11). The same is true for prints. The various dyes of plant, insect, or synthetic origin, namely, contain UV-absorbing pigments. Pigments of pomegranate fruit, cutch, lac dye-S, annatto, and indigo, for instance, absorb UV radiation starting at 275.5 nm, 279.5 nm, 279.5 nm, 295.5 nm, and 279.5 nm, respectively. Due to their UV-absorbing pigments, dyes increase the UPF of fabrics. Obviously, the UV range covered by the pigments, and their amount influences the obtained increase. Dying, for instance, cotton fabrics with high concentrations of *Lithospermum*

Erythrorhizon root, natural indigo (*Indigofera Indica*), annatto (*Bixa Orellana*), gardenia, sodium copper chlorophyll, or cochineal (*Dactylopius coccus*) yields about 80% UV absorption [20].

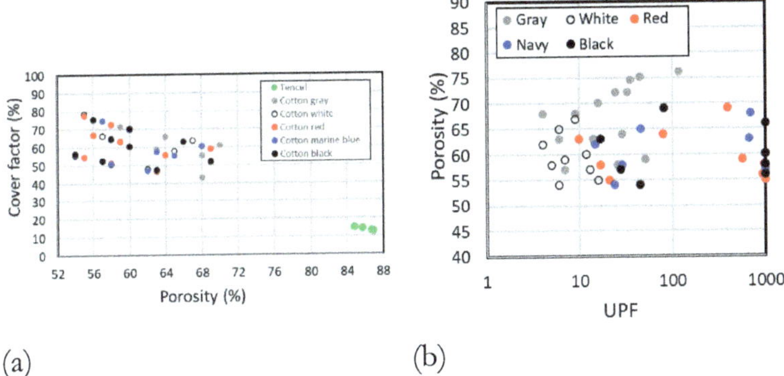

(a)   (b)

Fig. 2.11. Relation between (a) porosity and cover factor, and (b) UPF and porosity for the same fabrics shown in Fig. 2.8, but now identified by fabric color. Data from [21], [22], [23].

In contrast to natural dyes, synthetic dyes are not biodegradable and may lack UV-absorbing pigments. Consequently, producers of sun-protective clothing prefer natural dyes not only for environmental reasons. The UPF of wool fabric dyed with orange-peel extracts, for instance, is about six times higher than that of wool fabric dyed with a typical synthetic dye of similar shade and color depth [61].

In general, UPF increases with the depth of color, i.e., *color strength* (Fig. 2.12b). Color strength is the ratio of the *sorption*[25] coefficient, K, to scattering coefficient, S. It is given by the so-called *Kubelka-Munk equation* as

$$K/S = (1-R)^2/(2R)$$

Where R is the reflectance of the fabric color (Fig. 2.12a). Higher K/S values mean higher color strength. Recall, a photon hitting a surface can either be transmitted, reflected, or absorbed (Fig. 2.3b). Therefore, UV radiation entering a porous medium like a fabric can also be reflected within the pores. Even multiple scattering may occur in pores. The darker a color, the lower is its reflectance, and the higher is

---

[25] Sorption describes the attachment of substances to another.

the absorptance. Therefore, at the same porosity and cover factor, dark colors like indigo, navy, or black absorb more UV radiation, and, hence, provide higher sun protection than light or pastel colors. Furthermore, for same fabric material or blend, UPF increases linearly with the color strength (Fig. 2.12b).

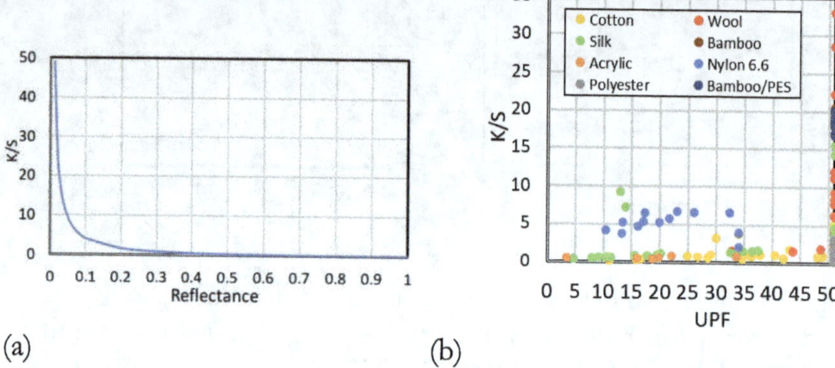

(a)   (b)

Fig. 2.12. Relationship between (a) color strength, K/S and reflectance according to the Kubelka-Munk equation, and (b) UPF and K/S. UPF values greater than 50 are shown at 51 on the right x-axis independent of their magnitude. Data from [20], [51], [52], [53], [55], [61], [62], [63].

2.3.1 The Impact of Dye Concentration on UPF

K/S values increase with increasing concentration of the colorant in the fabric (cf. Fig. 2.13). The color strength also increases with increasing time in and temperature of the color bath.

Obviously, synthetic fibers provide good UV protection (Fig. 2.13). Undyed polyester fabric can even provide excellent UV protection (UPF = 46.3) when a UV-absorbing agent is incorporated into the polyester during spinning. Therefore, despite dyeing polyester with natural dyes increases the UPF, color is of negligible importance for sun-protective polyester clothing [64]. The UPF values of the nylon 6.6, acrylic, polyester, cotton plain, twill, and satin weave blank samples (*greiges*) were 5.2, 3.4, 3.8, 19.2, and 13.3, respectively. Dyeing cotton twill or sateen with madder, indigo, and cochineal at 2, 4, and 6 omf% concentrations led to UPF values greater than 50 [20]. The plain, twill, and sateen cotton weaves had a thickness of 0.35, 0.69, and 0.61 mm with a fabric area weight of 120, 258, and 235 g/m^2, respectively. These results indicate that dyed "heavy" cotton fabrics are a better choice than thin and/or undyed cotton fabrics. Furthermore, natural colorants can yield high protection against UV radiation if the depth of color is a dark hue or the colorant concentration in the fabric is high (cf. Fig. 2.13; [20], [64]). This means you could take the weight and darkness of cotton fabrics as a broad indicator of the garment's sun-protection ability when no UPF information is available.

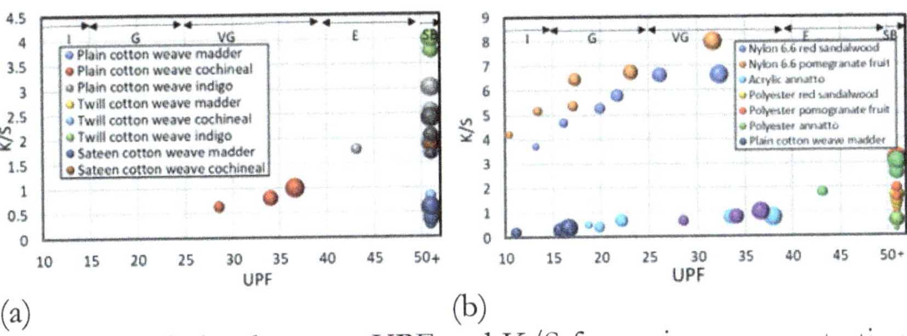

Fig. 2.13. Relation between UPF and K/S for various concentrations of dyes indicated by the size of the spheres for (a) cotton of different weave structure and color, and (b) different yarn materials and natural dyes. In the upper part, I, G, VG, E, and SB stand for insufficient, good, very good, excellent sun protection and sun-blocking. Spheres plotted right of 50 have UPF values exceeding 50. Data from [20], [53], [64].

When the path length through the fabric, i.e., fabric thickness, yarn and/or stitch count increases, the UPF increases as well. The pigments and waxes of wool fibers provide limited or no UV protection. The UV transmittance of undyed wool fabric, for instance, increases with increasing UV wavelength from 4% to 37% at 290 nm and 400 nm, respectively! As a result, undyed wool has a very low UPF (10.8). Dyeing wool fabrics with eucalyptus-leaf extract leads to a yellow-brownish shade, and reduces the UV transmittance to less than 5% in the UVB region [53]. The resulting UPF increases with increasing dye concentration from 32.8 (5 g/L) to 50.4 (20 g/L), meaning very good and excellent UV protection, respectively.

When dyeing silk with eucalyptus-leaf extract, UPF also increases with concentration (Fig. 2.14). However, the resulting pale yellowish-brown fails to achieve good UV protection [55]. On the contrary, dyeing non-mordanted *Eri* silk with natural dyes from Burma padauk bark, Andaman satinwood leaf, and neem bark results in UPF values of 201.94, 184.42, and 157.78, respectively [56]. This means not all dyes improve UV protection.

2.4 Enhancement of UPF with Mordants and Dyes

Mordants are substances, typically an inorganic oxide, that set the dye. Examples are ferrous sulphate ($FeSO_4$), copper chloride ($CuCl_2 \cdot 2H_2O$), zinc chloride ($ZnCl_2$), zirconium oxy chloride ($ZrOCl_2 \cdot 8H_2O$), aluminum chloride ($AlCl_3 \cdot 6H_2O$), and alum ($Al_2K_2(SO_4)$). Applying mordants in the color-application process improves color attachment and fastness. As a result, the UPF goes up; however, the degree of improved sun protection depends on the processes applied, dye and mordant types, their concentrations, and the raw material (e.g., Fig. 2.14). Mordanting undyed wool with aluminum-potassium sulfate ($AlK(SO_4)_2$), copper sulfate ($CuSO_4$), or ferrous sulfate ($FeSO_4$), for instance, reduces the UVB transmittance to less than 0.26%, 0.06%, and 0.05%, respectively. Like for dye, increasing the concentrations of the mordant (5 g/L, 10 g/L, 20 g/L) increases the UPF. The combined application of eucalyptus-leaf extract, and $FeSO_4$ mordant on wool, for instance, yields UPF values of 81.8 and 104.2 at dye concentrations of 5 g/L and 20 g/L, respectively [56]. Mordanting undyed *Eri* silk with $AlK(SO_4)_2$ increased the UPF from 35.08 to 49.9, while the UPF was 65.3 after mordanting with $FeSO_4$. The more effective UV protection after mordanting with $FeSO_4$ is due to the darker color achieved than with $AlK(SO_4)_2$ [56]. Dyeing silk with different concentrations of eucalyptus-leaf extract with different mordants of 10 g/L concentration improves UPF between good and excellent (Fig. 2.14). Independent of the dye concentration using a 10 g/L $FeSO_4$ mordant yields 50+ UPF values because of the dark grayish-brown color, and, hence, highest UV absorption [55].

Printing applies dye to only one side of the fabric. Because different colors yield different K/S and UPF (cf. Figs. 2.13, 2.14), we have to conclude that on print fabrics, UPF varies in space because of the different transmittance properties of the print dyes.

Obviously, obtained color strength differs among mordants and among the mordanting methods, even when using the same mordant (Fig. 2.14). For instance, when *pre-mordanting*, a cotton fabric dyed with mangrove bark extract achieves the highest color strength for $CuSO_4$, followed by $SnCl_2$, $FeSO_4$, and $AlK(SO_4)_2$. Dyeing cotton with mangrove bark extract without mordant leads to the lowest K/S value. When applying the dye and mordant concurrently (aka *meta-chrome* or *meta-mordanting*[26]), K/S decreases in the order $CuSO_4$, $FeSO_4$, $AlK(SO_4)_2$, and $SnCl_2$. In the *post-mordanting* (i.e., mordanting after dyeing) method, $FeSO_4$ followed by $CuSO_4$, $SnCl_2$, and $AlK(SO_4)_2$, provide the highest color strength.

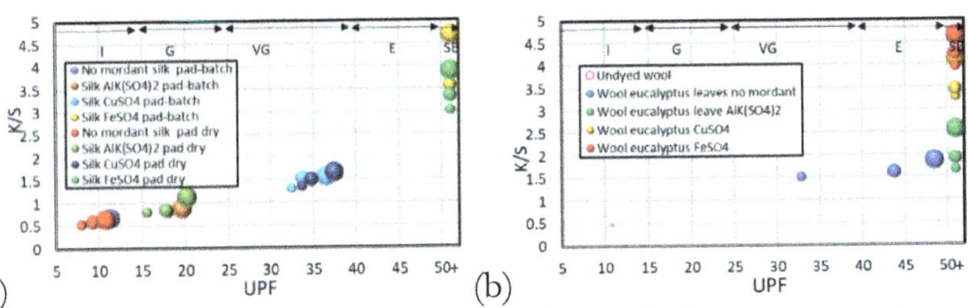

Fig. 2.14. Relation between UPF, K/S, and dye concentration of eucalyptus leaf extract (size of spheres) for (a) wool prior dyeing (greige), without and with various metallic mordants; and (b) plain-weave silk using two different pad-dyeing techniques, with and without 10 g/L of metal mordants. I, G, VG, E, and SB stand for insufficient, good, very good, excellent sun protection and sun-blocking. Data plotted to the right of 50 have UPF values exceeding 50. Data from [53], [55].

In general, ferrous sulfate and copper-sulfate mordants achieve better color strength than aluminum-potassium sulfate and stannous chloride, except for pre-mordanting with stannous chloride. The effect of mordanting on UPF indicates that increasing the time for dyeing results in more dye entering into the fiber, hence yielding darker shades. Darker shades provide higher UPF values in both the UVA and UVB range [52]. In contrast to dye, there exists a critical mordant concentration for achieving the maximum color strength. Once this critical concentration is exceeded, the color strength decreases (Fig. 2.15.).

Obviously, dyes that have their absorption maxima in the UV-region protect fabrics from photo-fading caused by UV radiation exposure. Concurrently, the presence of UV-absorbers insignificantly enhances the UV protection [65].

---

[26] Adding the mordant in the dye bath itself works only for certain dyes.

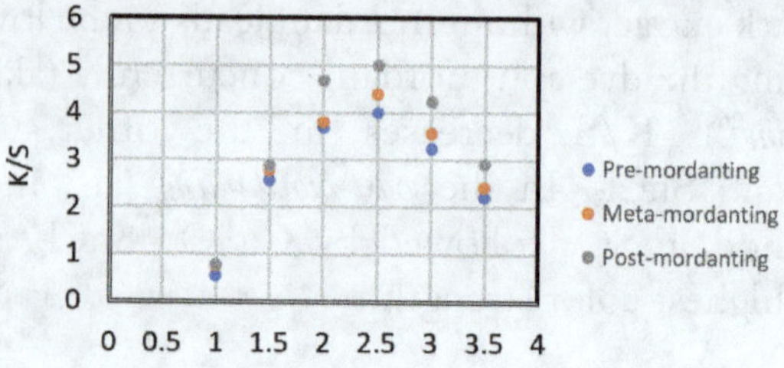

Fig. 2.15. Sensitivity of color strength to the concentration of $FeSO_4$ mordant and mordant method as obtained for cotton single-jersey fabrics dyed with green walnut-shell dye. In all cases, UPF exceeded 50+. Data from [52].

Unfortunately, details of the dyeing treatment and the color's location of maximum absorption in the solar spectrum remain non-disclosed to the consumer. Therefore, looking for UPF-labeled garments is your safest choice, followed by holding the garment against the Sun (If you can see the Sun through the fabric, it is not sun-safe.), darkness of the color, and fabric weight in the case of cotton.

2.5 Enhancement of UPF with Metal Oxides and Minerals

A recent development in textile production is to increase the UPF of fabrics with so-called *sun-bouncing minerals*. Technically, these minerals are UV-absorbing metal oxides. This method applies titanium dioxide (TiO_2) or zinc oxide (ZnO) into the yarns or onto the fabric surface prior to dyeing. These metal oxides are crystalline solids containing an oxide anion and metal cation. Their reaction with water forms bases, while their reaction with acids forms salts.

The color of natural fibers like cotton or wool is typically not pure white. Therefore, applying dye and creating a uniform color is difficult. Consequently, natural fibers are often bleached prior to dyeing. Bleach can be *reductive* or *oxidative*. Oxidative bleaches like hydrogen peroxide (H_2O_2) break the chemical bonds with the pigments. As a result, the chemical nature changes, losing the ability to absorb light. Reductive bleaches like sulfur dioxide and sodium bisulfite ($NaHSO_3$) alter the double bonds of pigments to single bonds. As a consequence, light is absorbed. Sodium hypochlorite ($NaClO$), for instance, reacts to metal oxides.

When looking at half-bleached cotton knits, for instance, UPF increases with increasing fabric weight, thickness, and decreasing porosity. Recall, fabric weight and thickness depend on the knit or weave structure. Therefore, the UPF of interlock fabric is greater than for piqué followed by parasol fabric (Fig. 2.10). Pre-mordanting these cotton knits with metal salts improves both the K/S- and UPF values. For the same fabric structure and natural dye, the improvement is highest for Zr-oxychloride, followed by Cu-chloride, Zn-chloride, Al-chloride, and no pre-mordant [51]. These pre-mordants deposit metal oxides onto the fabric/yarn when reacting with reducing bleach. Dyeing these pre-mordanted fabrics/yarns with natural dyes notably improves both the K/S- and UPF values. The degree of improvement depends on the physical and chemical characteristics, concentration of the natural dye, type of mordant, and fabric structure [51]. Luckily, sweat, chlorine, sea salt, and multiple laundering do not impair the sun-protective clothes when UV-absorbers are infused at the fiber level.

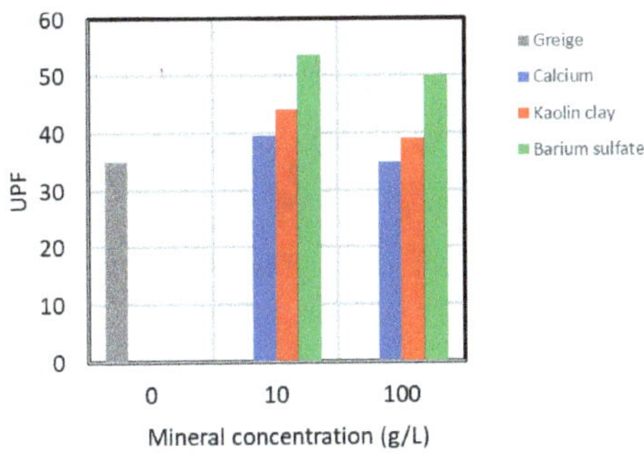

Fig. 2.16. UPF of brown reactive-dyed, plated single-jersey fabric with 91/9 cotton/spandex fiber content without (greige) and with different mineral coatings at 10 and 100 g/L concentrations. Data from [66]

2.6 Conclusions on Sun-Protective Fashion

What makes clothes sun-safe depends on a very complex relationship between fabric weight, stitch density, knit or weave patterns, yarn treatment with UV-absorbers or nanoparticles, and dye. Fabric weight affects UV protection the most, while its thickness and stitch density play a minor role.

Unless you wear UPF 50+ clothing, some of these UV rays may travel through your clothes and reach your skin. In simple words, you might even get a sunburn when fully dressed. Therefore, for the sake of your skin health, you should use skin-care products with sunscreen all the time. When you spend a lot of time outside, protect your skin with UPF-rated clothing, even when staying in the shadow. Today sun-protective clothing comes in very chic and even professional cuts that don't read sparse time, great outdoors, or vacation.

In general, dark (e.g., black, dark blue) or bright, vivid colors (e.g., red, yellow) absorb more UV radiation than light colors (e.g., pastels, white). As a result, less UV radiation reaches your skin when choosing dark or bright colors.

Typically, clothing tags don't disclose details about the color application process. Therefore, opting for sun-safe clothing and applying sunscreen to uncovered skin are your best measures of protection. Consequently, make wearing sun-protective clothes and sunscreen a habit.

Regarding sun exposure, be aware that more UV reaches the ground in pristine than polluted air. Prefer cloudy over cloud-free for being outside. Also, reduce the time that you spend outside around local noon if you can. Take your walk, run, or shopping trip early in the morning or in the evening. Ask yourself whether the great view from that mountain is really worth it. If not, hike in the valley or at a lower altitude.

3 THERMO-PHYSIOLOGICAL COMFORT OF FABRICS

3.1 Staying Warm or Cool – The Role of Conduction and Convection

Conduction refers to the transfer of heat and moisture by molecules. A well-familiar example of thermal conduction is an iron pan on a hot stove. Despite the iron handle having no contact with the stove, it gets hot because the iron molecules transfer heat to the iron handle. Such a conduction process also occurs between different molecules and over the border of different materials.

Because our body consists of molecules, it can lose heat due to conduction in many ways. For instance, in winter, heat from your feet is transferred to the sole of your boots and then to the frigid cold ground. The result is that you lose body heat. Consequently, your body has to compensate for the lost energy by turning chemical energy (aka food) into heat. Whenever your body temperature exceeds the ambient air temperature, your skin conducts heat to the air between your clothes and skin or, in the case of uncovered skin, from your skin to the ambient air. On the contrary, when the ambient temperature exceeds your body temperature, you gain heat and feel hot.

Wherever air is in direct contact with your skin, conduction takes place. Clothing hinders this energy loss by acting as an insulator. An important property of any insulting material - including clothing - is the product of its *specific heat capacity* and *density*. The specific heat capacity expresses the material's ability to hold heat. Translated into terms of fashion: The higher a fabric's specific heat capacity, the longer it takes for the fabric to cool down to the ambient air temperature when you go outside into the cold.

The lower the thermal conduction, the better the insulation. In case of air, thermal conduction is lower for thin air layers than thick layers because in thick layers *convection* and movement occur. Convection refers to the transfer of heat from one place to another within a gas (e.g., air) or liquid (e.g., water). The *heat-transfer rate* by convection reads

$$Q_{conv} = -H \cdot A \cdot (T_s - T_a)$$

Where Q_{conv} is the *convective heat flow*, T_s is the temperature of the material (e.g., your skin), and T_a is the ambient air temperature. Furthermore, H is the *convective heat-transfer coefficient*, and A is the (cross-sectional) area through which the heat flow occurs.

What does this mean for you? When you think about it, even indoors, there is a huge difference between your skin- and the room temperature. Consequently, your skin increases the air temperature in its immediate vicinity via conduction. Because, at the same moisture and pressure, warm air is lighter than cold air, the warmed near-skin air becomes buoyant and rises. This rising of buoyant air is called convection. Convection transports heat away from your naked skin.

Table 3.1. Thermal properties of various raw materials[27] (i.e., not fibers, nor fabrics), water, soil, and air. Data from [16], [67], [68], [69].

Material	Specific heat capacity $(J/(kg\,K))$	Bulk density (kg/m^3)	Thermal conductivity $(W/(m{\cdot}K))$
Water	4180	1000	0.58
Air	1004	~1	0.024
Cotton	1162	1540-1560	0.06
Bamboo	1800	4510-7800	0.188
Flax	2540	1280–1600	0.038-0.075
Polyester	750-1316	1400	0.05-0.24
Nylon	1500	1150	0.2
Leather	1500	1200–1300	0.13-0.14
Silk	1380	1100	0.066-0.122
Loose Wool	1260	1341	0.094-0.12
Rubber	1130-4100	910-930	0.14-0.15
Human body	2440-3390	980	0.3
Soil	750-760	2500-2660	3.0-8.8
Organic soil	1900	1300	0.3

In fabrics themselves, however, very fine fibers damp out all convection. The fine fibers' huge surface area namely impedes the free flow of air past the fibers.

---

[27] Specific heat capacity, bulk density, and conductivity of air have a marginal, but negligible sensitivity to temperature over the typical range of air temperatures.

Besides this natural rising of heated air, convection may be forced by external pressure differences, such as wind or body motion. In this case, we speak of *forced convection*[28]. A well-familiar example of forced convection is a mom's blowing on her kid's food or burned skin to cool it (Fig. 3.1b).

In the case of clothes, forced convection can be due to wind or ventilation caused by the wearer's motion. At temperatures less than body temperature, the wind blows the slightly warmed near-skin air away from your skin thereby accelerating your skin's energy loss. Similarly, when you move, your motion forces the air between your skin and clothes to move (Fig. 3.1c).

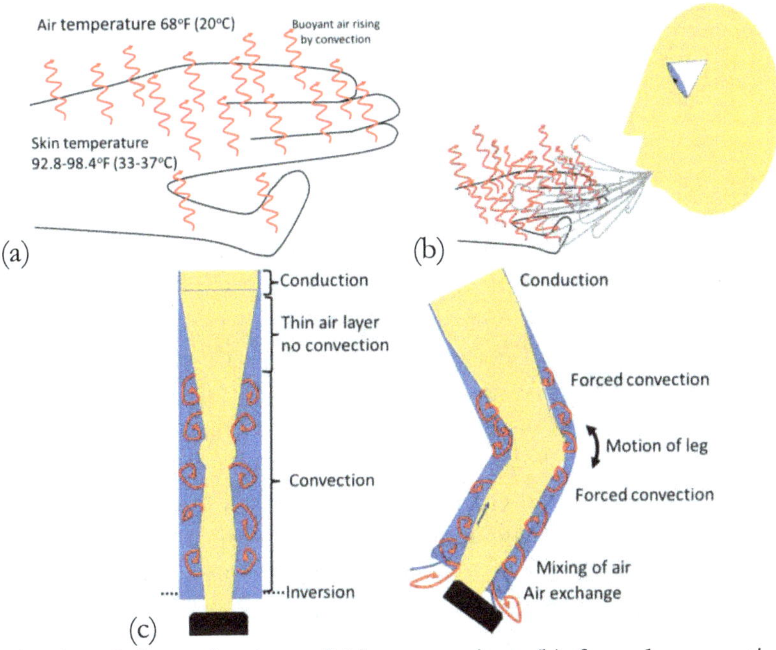

Fig. 3.1. Schematic view of (a) convection, (b) forced convection, (c) conduction, and convection while standing still, and conduction, forced convection, and air exchange while walking. Red arrows symbolize the rise of warmed near-skin air.

---

[28] In general, forced convection refers to a process in which gases or liquids are forced to move to increase the heat transfer. The forcing can be due to a fan, pump, suction device, shear, wind, etc.

When standing still, warm air is above cold air building an *inversion* (Fig. 3.1c) that suppresses air exchange. However, the forced convection from walking breaks the inversion layer leading to the mixing of cold ambient air with the slightly warmed air inside your clothes.

3.1.1 Thermal Resistance and Thermal Conductivity

Thermal conductivity is a material property (e.g., Tab. 3.1). It measures the amount of heat transferred from a unit thickness of material to a unit surface area under *steady-state* (i.e., not changing) conditions when the heat transfer only depends on a temperature difference. In simple words, it measures how fast the material conducts heat.

According to *Fourier's law*, we can calculate the heat flux due to conduction, Q_{cond}, as:

$$Q_{cond} = -k \ A \ dT/dx$$

Where k is the *thermal conductivity coefficient*, A is the (cross-sectional) area through which the thermal energy passes and dT is the temperature difference over a given distance, dx, of material.

Thermal conductivity gives the insulating capability of any material. It quantifies the heat flux that occurs through the fabric material for a given temperature *gradient*[29] between the top and bottom side of the fabric. The distance, dx, between the two sides is the fabric thickness, h.

Thermal insulation in the form of clothes reduces the heat transfer between your body and the environment because it decreases the thermal heat conduction. Low thermal conductivity means that the material has high insulating capability, and is a good insulator. Therefore, fabrics with low thermal conduction are great to stay in thermal comfort outside for an extended amount of time. A down coat, for instance, can be a great insulator. Because thin layers of air are bad thermal conductors (cf. Tab. 3.1), layering of clothes can help to create appropriate insulation to stay outside for several hours.

Thermal resistance indicates how well a fabric insulates your body from the ambient air. It depends on the fabric's thickness and thermal

---

[29] In general, $d\chi/dx$ gives the change of any quantity χ in the direction x, and is referred to as the gradient of χ.

conductivity. Thermal resistance is the ratio of the temperature difference between the two faces of the fabric to the rate of heat flow per unit area through the fabric. In simple words, the thermal resistance, R_{ct}, of a fabric is the fabric's resistance to heat flux; it is proportional to the fabric thickness, h. Under idealized conditions, this relation reads

$$R_{ct}=h/Q_{cond.}$$

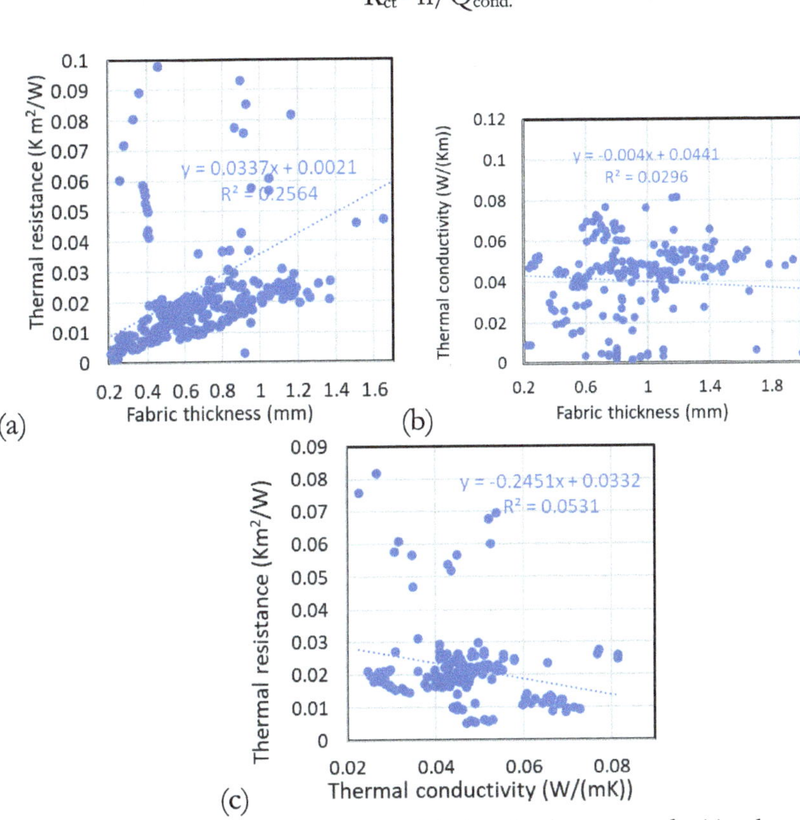

Fig. 3.2. Relationship between fabric thickness and (a) thermal conductivity, (b) thermal resistance, and (c) relation between thermal conductivity and thermal resistance. Equations describe the relationship (visualized by the dotted lines) between the quantity on the x- and y-axis. R^2 multiplied by 100 is the percent of correlation among the values on the x- and y-axis. Data from [14], [15], [26], [27], [29], [30], [31], [32], [33], [34], [35], [36], [38], [70], [71], [72], [73], [74], [75], [76], [77], [78], [79], [80], [81].

Because thermal conductivity decreases as thermal resistance increases, we can say that thermal resistance increases as fabric thickness decreases (Fig. 3.2). Obviously, thermal resistance and thermal conductivity are important parameters for thermal comfort in both cold and hot weather dressing.

3.1.2 The Role of Fabric Material Porosity for Air Permeability, Diffusion, Thermal Conductivity, and Thermal Resistance

Thermal conductivity and thermal resistance are material properties. Therefore, their values differ among fabrics of different materials including air (e.g., Fig. 3.3). Because the thermal resistance of air is relatively high as compared to textile fibers, fabrics with many small air pockets (high porosity) provide better insulation than those with large spaces between yarns (loose knits/weaves) or tight weaves with low porosity[30]. Consequently, the increased thermal resistance of the fabric (and decreased fabric density) reduces the heat transfer (Fig. 3.3).

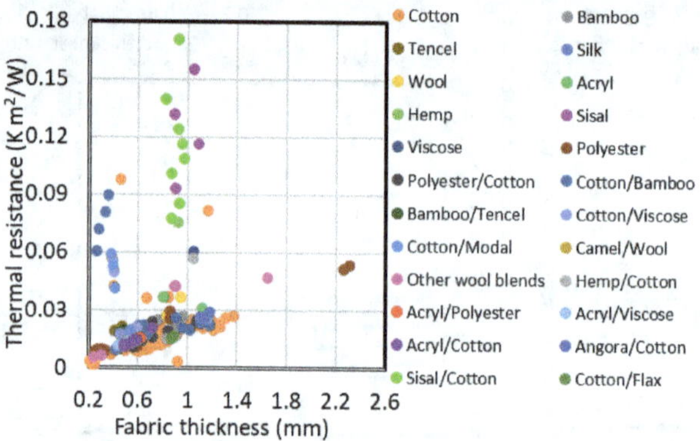

Fig. 3.3. Thermal resistance of various fabric materials vs. fabric thickness. Modified after [82]. Data from [10], [14], [15], [26], [27], [29], [30], [32], [33], [34], [35], [36], [37], [38], [39], [62], [71], [72], [73], [74], [75], [76], [77], [78], [79], [81], [83], [84], [85], [86], [87], [88], [89].

In textiles, *thermal diffusion* refers to the heat flow through the air within the fabric structure. Thermal diffusivity is their temporary thermal characteristic. In general, the *air permeability* of fabrics depends

---

[30] Recall that repeated brushing and cooking of fabrics served to create loden, melton, and duffle for protection from cold, windy weather.

on external ambient conditions (e.g., temperature- and pressure differences between your body and the environment), the physical properties of the material (e.g., water vapor, air) passing through the fabric, and the fabric porosity (Fig. 3.3). As explained in Ch. 2, fabric porosity depends on the structural parameters of the fabric like yarn twist, roughness, yarn count, material, the weave or knitting pattern, and for blends, the mix ratio.

In the textile industry, porosity at the fabric scale refers to the space between fibers (Fig. 2.8). However, the fibers themselves can be porous too. Angora hair, for instance, is hollow. These pores within the fiber enclose air. Because air is a bad conductor (Tab. 3.1), these air-filled pores influence thermal conductivity and resistance and affect fabric density (cf. Fig. 2.9).

Typically, low fabric density means high air permeability (Fig. 3.4a). Furthermore, in general, a high number of pores per unit volume means a low air permeability (Fig. 3.4b). This counter-intuitive behavior is because the *drag force* to airflow increases with increasing surface area. Let me explain this to you. For simplicity, let's assume the pores are cylinder-like tubes. Then the surface area of a pore is that of a cylinder $A=2\pi rh+2\pi r^2$ where h is the height or in the case of the pores, the path length. When the pores are open at their two ends, the pore surface area is $A=2\pi rh$. Because the same volume of fabric (given by the product of unit area times thickness) can hold many more small pores than large pores, the sum of the surface areas of the many small pores exceeds that of the fewer large pores. Consequently, at the same thickness, fabrics with fine yarns have more small pores than those with coarse yarns. Therefore, we can say that for fabrics consisting of just one material, air permeability decreases with increasing fiber fineness.

The thermal conductivity of fabrics made from the same fibers depends on fabric density (cf. Fig. 3.4c). Interestingly, thermal conductivity remains in the same range independent of the air permeability for the fabric of the same material and material blends (cf. Fig. 3.4d). Therefore, you don't have to compromise low thermal conductivity for high air permeability. Because air has low conductivity, fabrics with high porosity have low conductivity (Fig. 3.4e).

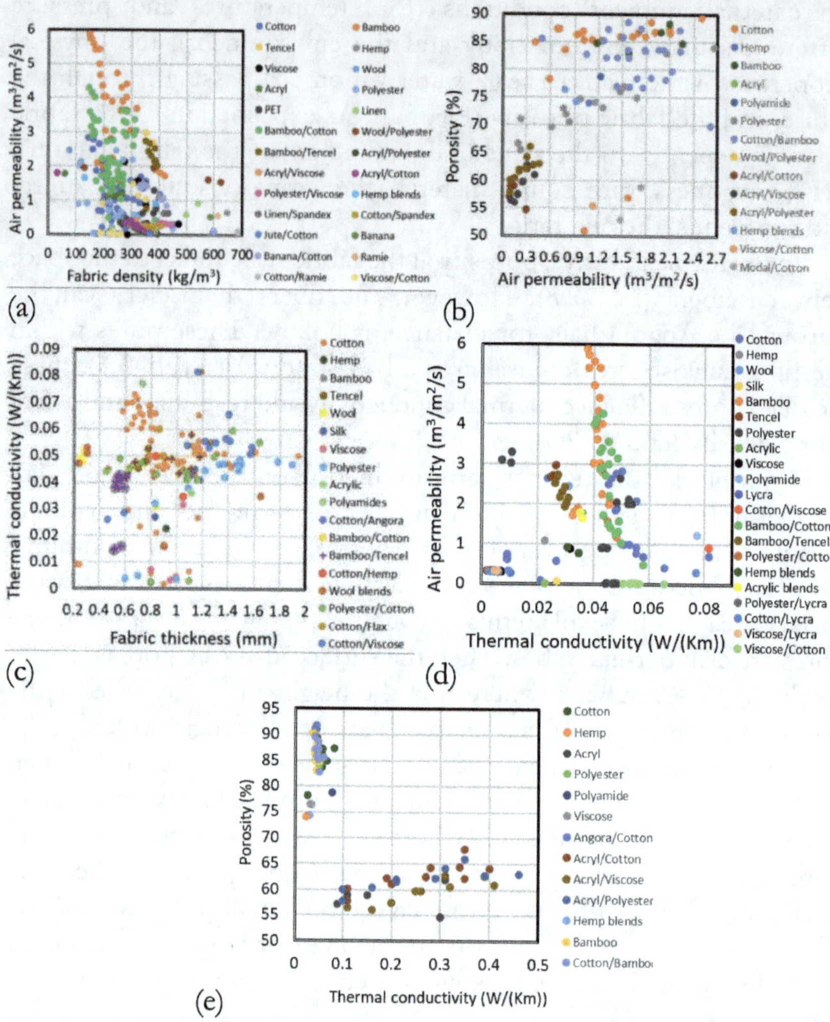

Fig. 3.4. Relation between (a) air permeability and fabric density, (b) air permeability and porosity, (c) thermal conductivity and fabric thickness, (d) thermal conductivity and air permeability, (e) thermal conductivity and porosity for various fabric materials and blends. Color coding differs among panels. Modified after [82]. Data from [26], [29], [30], [31], [32], [34], [36], [37], [44], [71], [73], [74], [75], [78], [79], [80], [81], [90], [91].

Generally, fabrics with thermal resistance values below 0.055 m^2K/W are very comfortable. Increasing fabric porosity or fiber fineness increases thermal resistance (Fig. 3.4b). The fabric becomes a better insulator and keeps you warm longer than fabric with coarse yarns and large pores. While for viscose, Tencel, and bamboo thermal resistance more or less ranges between 0.01 and 0.3 m²K/W, air permeability seems to differ strongly for the same fabric material (Fig. 3.4a).

The blending of materials affects fabric thickness, weight, density, porosity, air permeability, and thermal resistance (Fig. 3.5). For instance, at the same stitch length, different Lycra contents increase the fabric weight of cotton/Lycra blends. However, at the same Lycra content, but different stitch length, fabric weight remains the same. This means that Lycra content has a higher impact on fabric weight than stitch length [92].

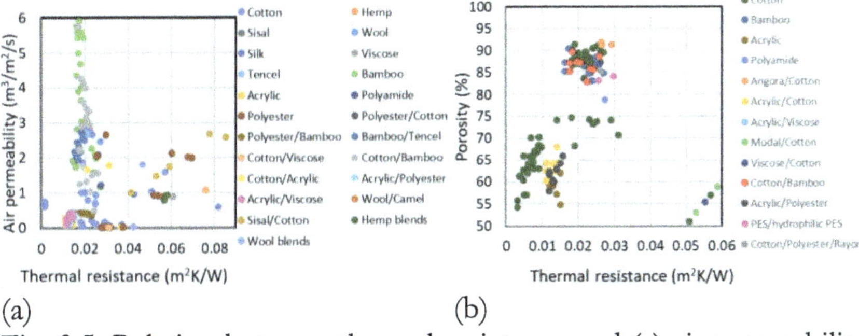

(a)   (b)

Fig. 3.5. Relation between thermal resistance and (a) air permeability, and (b) porosity for various fabric materials and blends. Color coding differs among panels. Modified after [82]. Data from [14], [26], [30], [33], [34], [35], [36], [38], [39], [70], [79], [80].

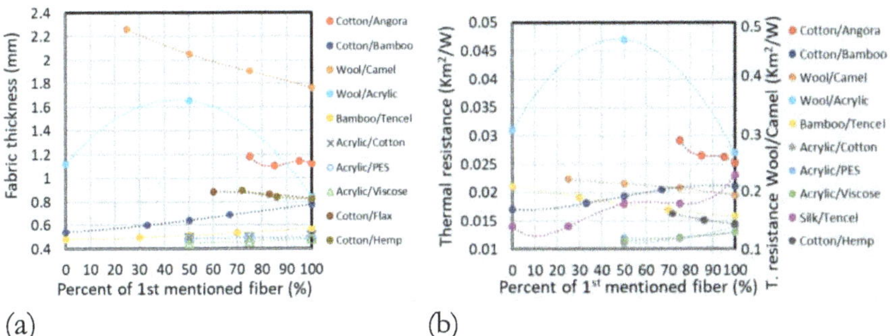

(a)   (b)

Fig. 3.6. Increasing the content of the first mentioned fiber in the blend vs. (a) fabric thickness, and (b) thermal resistance. Colors differ among panels. Modified after [82]. Data from [14], [29], [31], [33], [74], [84].

Blending cotton with nylon 6.6 reduces the fabric weight and increases air permeability as compared to 100% cotton [93]. Fabrics with high content of cashmere have high air permeability. Therefore, they pass body heat through at a higher rate than fabrics with low cashmere content [68].

The blend ratio affects thermal resistance non-linearly and at different rates depending on the blend ratio (Fig. 3.6).

3.1.3 The Role of Fabric Tightness/Density for Thermal Comfort

Because fabric structure (in terms of porosity) influences thermal resistance and thermal conductivity, let's take again a look at knitted and woven fabric structures. As explained in Ch. 1, the traditional Fair Isle and Lice knitting techniques catch unused strands every 3 to 5 stitches. Thereby, a lot of thin air pockets are created that increase insulation. Interestingly, the Fair Isle technique limits the increased insulation to the upper chest, while the Lice pattern increases insulation over the entire garment. Double-knit or bi-layer knit structures like, for instance, interlock typically have greater mean thickness and fabric weight per unit area than fabrics from the same yarn in single-knit structure.

Let's compare fabrics close in density (Fig. 3.7) to assess the impact of fabric structure on air permeability, thermal resistance, thermal conductivity, and porosity. In Figure 3.7, data for single-jersey represent different loop lengths, yarn thickness, tightness of knit, and the largest variety of raw materials. Therefore, thermal conductivity and thermal resistance of single-jersey highly vary with fabric density, and porosity-density relations arrange accordingly. Nevertheless, we can see that porosity, air permeability, and thermal resistance decrease with increasing fabric density. Thermal conductivity increases as fabric density increases.

Waffle-knit has high porosity (96.4%). Because the air permeability of fabrics increases with increasing fabric porosity (Fig. 3.4b), waffle-knit polyester has greater air permeability than polyester with other knit structures (Fig. 3.8); single-jersey polyester fabric has the lowest air permeability, and porosity (89%) [24].

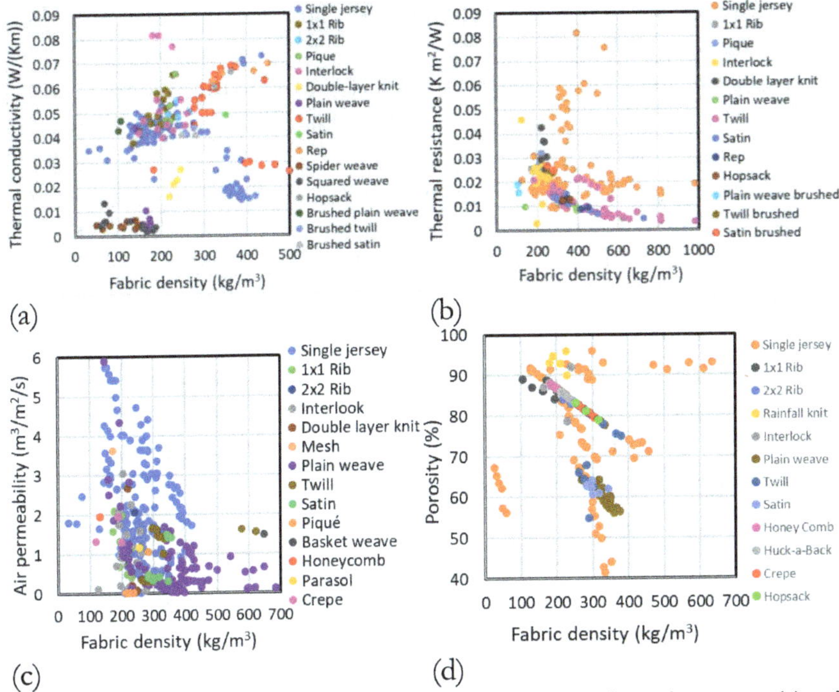

(a)

(b)

(c)

(d)

Fig. 3.7. (a) Thermal conductivity, (b) thermal resistance, (c) air permeability, and (d) porosity of various knits and weaves dependent on fabric density. Color code differs among panels. Data were not available for all knits and weaves. Modified after [82]. Data from [14], [15], [25], [26], [27], [29], [30], [31], [32], [33], [36], [38], [39], [40], [42], [54], [71], [74], [75], [76], [79], [81], [83], [84], [94].

Except for thermal resistance, cotton piqué shows the lowest thermal properties followed by cotton plain weave and sateen [94]. The thermal resistance of piqué is an order of magnitude higher than that of satin or plain weave (Fig. 3.8). While thermal resistance linearly decreases with increasing density for piqué, it remains almost constant for satin- and plain fabrics.

Plain and rep 2/2 cotton fabrics have the highest specific mass, and thermal conductivity (Fig. 3.7). In other words, their unit volume contains the most fibrous material through which thermal conduction occurs. The loose structure of cotton twill 3/1 fabrics entraps a lot of air inside. As a result, it has a relatively high thermal conductivity [27]. Thermal conductivity and thermal *absorptivity* of plain cotton fabric exceed those of 3/1 twill, 2/2 twill, 2/2 rep, 1/1 rep, and 2/2 hopsack fabrics of the same yarns. This means plain fabrics have a lower thermal resistance than the latter.

At the same fabric area weight (162 g/m^2) and blend (60/40 wool/polyester), basket-, twill-, and break-twill weaves differ in thickness (0.25, 0.26, 0.28 mm), and porosity (16.8, 13.3, 10.9%). Basketweave has a neat, dense structure, while twill has the finest yarns, i.e., the highest yarn count, but loosest structure. The 60/40 wool/polyester break-twill fabric has the highest thickness and thermal resistance followed by basket weave [40].

Air permeability decreases when fabric thickness increases. This means air permeability decreases when porosity increases [40] (Fig. 3.4). Consequently, the air permeability of the knitted fabrics can decrease with increasing yarn count and thickness. Interestingly, knits of the same tightness (stitch length) can have quite different air exchange depending on the yarn material. Some materials form yarns with filaments that can interlock with each other. Often even the mix of different materials in the spinning process leads to "closing" gaps formed by the knit.

The data reveal that for cotton and polyester fabrics thermal conductivity is notably higher for interlock and 1×1 rib than for single-jersey knit (cf. Figs. 3.3, 3.6). The reason is air entrapped in the fabric structure. Because the amount of fiber per unit area of fabric weight increases, the amount of air decreases. Recall that the thermal conductivity of air is lower than that of fibers. As a result, thermal conductivity is higher for heavier fabrics with less still air, for instance, interlock than for less heavy fabrics. Furthermore, a decrease in rib number leads to a decrease in heat loss. This means that, for instance, garments in 3×3, or 2×2 rib, provide less insulation than a garment in 1×1 rib (Fig. 3.7).

In general, the thermal conductivity of cotton, wool, and fleece are 0.168 - 0.184, 0.04 - 0.07, and 0.035 W/(m·K), respectively. Consequently, you stay warm longer in fleece than in wool than in cotton. Of course, the thickness and tightness of fabric play a role. A single-jersey cotton fabric with a unit area mass of 1 g/m^2 (0.23 oz per square inch) has a thermal conductivity of about 0.035 W/(m·K). The thermal conductivity of interlock and 1×1 rib cotton fabrics exceeds that of cotton jerseys. The reasons are that still air is a bad conductor and when the fabric gets tighter the space and amount of "air pores" goes down. The thermal conductivity of a polyester single-jersey at the same density as that of the cotton jersey is about 0.032 W/(m·K). Therefore, you are better off with a polyester sweater in cold climates than with a cotton sweater of the same tightness of knit. Again, when the knit gets tighter the thermal conductivity goes up because the pore space for still air goes down.

Together these studies showed that fabric structure significantly affects the porosity, air permeability, and thermal resistance or conductivity of fabrics. The pattern of yarn-interlacing points is namely specific for each weave or knit structure. Low numbers of yarn-interlacing points mean high porosity, air permeability, thermal conductivity, and low thermal resistance [35].

3.1.4 Thermal Properties of Fabrics from Blended Yarns or Blended Weaves

The blending of fibers affects their *energetic properties* (thermal conductivity, thermal resistance, porosity, air permeability). Recall, in blended fabrics, yarns may have different filaments and/or fineness. Consequently, the filaments may close pores, especially, when the blend is made at the fabric manufacturing level. Clearer pores mean less surface area, and, hence, increased air permeability.

Figure 3.6 shows that hemp or hemp-blended knit fabrics exhibit similar thermal characteristics as cotton and viscose fabrics. Cotton/hemp blended textiles show reduced air (and water-vapor) permeability with the downward trend in thermal resistance. The degree to which the transport properties vary among the cotton/hemp knits depends on the twist intensity of the cotton yarns [30].

Air permeability is the lowest for 100% wool and 97.5/2.5 wool/cashmere fabrics, while it is highest for 80/20 followed by the 90/10 and 95/5 wool/cashmere fabrics [68]. A major reason is the microscopically smoother, and much thinner cashmere-fiber diameter as compared to wool fibers. Consequently, worsted fabrics with high cashmere content have clearer pores than wool or wool blends with low cashmere content.

Inserting, for instance, 10% or more cashmere fiber in worsted fabrics increases their thermal resistance. While the thermal conductivity of mixed worsted fabrics differs notably, no specific trend exists for the mix due to the different thicknesses of these fabrics. Thermal conductivity for 90/10 wool/cashmere fabrics has the highest thermal conductivity of the wool/cashmere blends. This means cold weather clothing benefits from the use of cashmere fibers.

Let's take a look at other fiber blends. In the case of Tencel/cotton and Tencel/bamboo blends, thermal conductivity and fabric thickness decrease with increasing fraction of Tencel fiber. Air (and water-vapor) permeability increase with increasing Tencel content [29]. Because bamboo yarns are less hairy than cotton yarns, Tencel/bamboo blended fabrics have generally a lower porosity than cotton or Tencel/cotton fabrics (Fig. 2.9).

When looking at polyester and polyester blends, we find the following: Fabric from 100% polyester fibers has maximum air permeability due to the regular cross-section of the polyester yarn. Air permeability decreases with increasing fabric thickness (Fig. 3.4b) and weight. For the same reason, blending cotton with nylon 6.6 reduces the fabric weight and increases air permeability as compared to 100% cotton [93].

Yarn configuration is not only an important factor for UV protection by multifilament woven polyethylene terephthalate (PET) fabrics (Ch. 2). Fabric from flat yarn without twist has the least air permeability. The very compact structure of twisted PET yarns leaves maximum porosity for air permeability when used for twisted warp and weft fabrics [54].

Overall, we can say that thermal conductivity and thermal resistance depend on fabric structure, raw material, and the ratio of the blend (if applicable). Thermal conductivity increases with fabric thickness, while thermal resistance decreases with increasing fabric thickness.

3.1.5 Love at First Touch - Thermal Absorptivity

When you touch a fabric that has a different temperature than your skin, your hand and the fabric exchange heat. *Thermal absorptivity* is a measure to quantify this heat exchange. In simple words, it gives the warm or cool feeling of a textile on the human skin upon the first sensation. Fabrics with high thermal absorptivity feel cooler at first contact than those with low thermal absorptivity.

Smooth fabrics like, for instance, satin (Fig. 3.8) have a large surface area between the fabric and skin, which causes a cool feeling. A large surface area also means an increased area where evaporation of sweat may occur. On the contrary, fabrics with rough surfaces (e.g., brushed fabrics) provide only low skin contact on the first touch. Consequently, the heat exchange is small, and we experience a warm sensation. In simple words, the smoother the fabric structure, the cooler it feels.

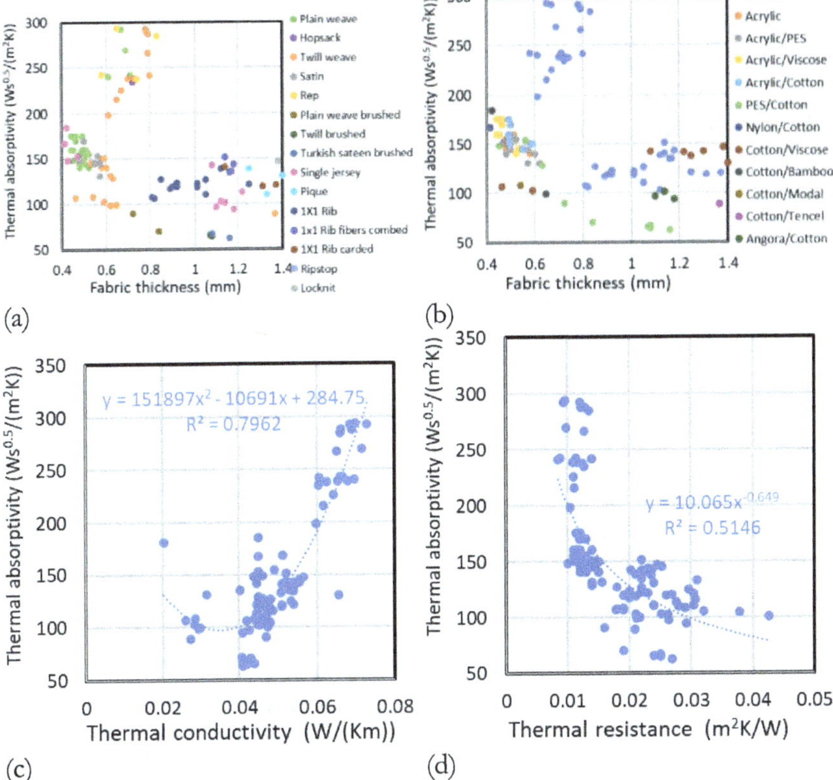

Fig. 3.8. Thermal absorptivity for various (a) fabric structures, (b) fabric materials, and the relation between (c) thermal conductivity and thermal absorptivity, and (d) thermal resistance and thermal absorptivity. Data from [14], [15], [27], [33], [34], [70], [72], [79].

A comparison of cotton rib and cotton twill (Fig. 3.8) suggests that fabric material is of less importance for thermal absorptivity than fabric structure. When looking at fabrics close in thickness (Fig. 3.8a, b), you see the effect of the combination of fabric structure and material on thermal absorptivity. Thermal absorption increases from piqué (warm feeling) over plain weave to satin (cool feeling). For all three weaves thermal conductivity linearly increases with increasing density [94]. Furthermore, thermal conductivity, and thermal absorptivity increase with decreasing thermal resistance (cf. Fig. 3.8c, d). Consequently, satin and brushed fabrics are best for bedding or clothing in summer and winter, respectively.

In a nutshell, fabric tightness influences thermal conductivity, resistance, and absorptivity as well as air permeability notably. Loose fabrics provide higher insulation and air permeability values as well as a warmer feeling than tightly knitted or woven fabrics.

3.2 Moisture Management of Clothes

When a design requires printing or dyeing of fabrics, *hydrophobicity* of the fibers, *sorption*, liquid uptake rates, liquid transport, *spreading*, and *wicking* are properties to consider in the choice of fabric. These aspects are also important for wearing comfort. Physical activity or high ambient temperatures require to release excessive heat by sweating to maintain a stable body temperature. When we sweat, the sweat is a moisture flux by diffusion through the skin given by

$$LE_{skin} = 4.0 + 0.12\,(e_{skin} - e)$$

Where LE_{skin} is the *evaporative heat flux* (so called *latent heat flux density*) in W/m^2, e_{skin} is the *saturation water-vapor pressure* at skin temperature, and e is ambient *water-vapor pressure* (both in hPa or mb).

Permeability, in general, refers to the rate of gas or liquid flowing perpendicularly through the fabric. A material can be permeable for some gases, while being impermeable for other gases. Therefore, while air may contain water vapor, high air permeability of a fabric does not guarantee high permeability for water vapor[31], and high *water-vapor transmission rates* (WVTR). Moreover, high porosity does not lead to high WVTR (Fig. 3.9). In textiles, often the term *breathability* serves to describe the *water-vapor permeability* of a garment.

Like air permeability, we can express water-vapor or *moisture permeability* as the speed of water-vapor or sweat passing through a fabric under a prescribed pressure difference between the two fabric-surface sides. While the fabric material doesn't influence air permeability, it affects water-vapor- and moisture permeability.

---

[31] A fabric acts like a filter. Only material smaller than the pores can pass through.

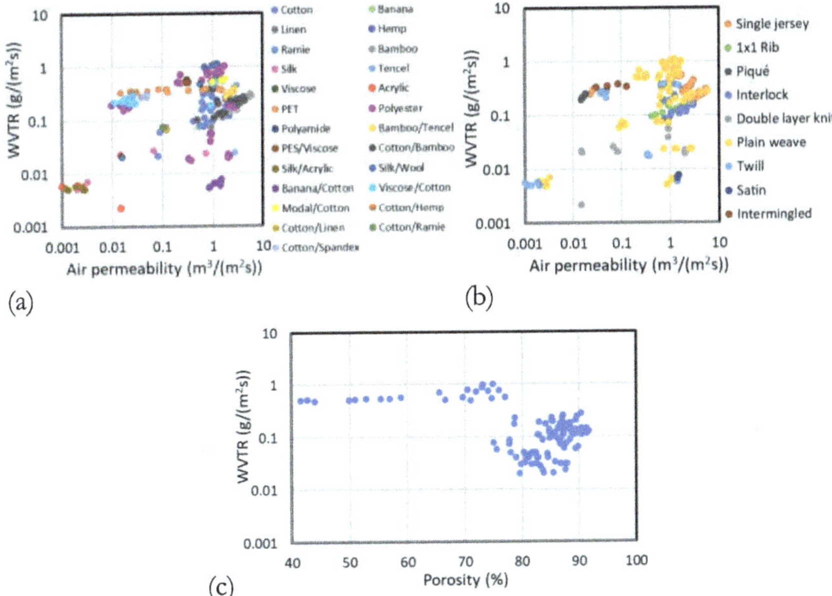

Fig. 3.9. Air permeability vs. water-vapor transmission rate (WVTR) for fabrics of various (a) fiber types, and (b) structures. (c) Relationship of WVTR and porosity. Note that the x- and y-axes in (a) and (b), and the y-axis in (c) are logarithmic. PET and PES are polyethylene and polyester. Data from [25], [26], [77], [78], [79], [95], [96], [97], [98].

The reason is as follows. Air permeability occurs only by diffusion. On the contrary, four processes govern the moisture and water-vapor transport through fibrous substrates. As part of the air, water vapor passes through the spaces between fibers by diffusion too. However, the fibers can absorb, transmit and then desorb water vapor. Forced convection can migrate water-vapor transfer because, at the same temperature and pressure, water vapor is lighter than dry air. Furthermore, water can be adsorbed and transmitted along the fiber surfaces. Consequently, key fabric parameters that influence a person's perception of moisture comfort are water-vapor- and water (sweat) permeability, porosity including surface area, and *wettability*.

3.2.1 Definition of Moisture Permeability, Moisture Transfer, and Absorption

Diffusion through the fabric pores makes up almost all of the water-vapor transfer from the skin to the ambient air or vice versa depending on where the water vapor (or relative humidity) is lowest.

Unfortunately, due to differences among disciplines as well as regional units, terms and units for water-vapor permeability differ which prohibits direct comparisons. Therefore, let me explain the most often used terms briefly before we will use the terms normalized [82] to the same units later to compare the various researchers' findings.

Water-vapor resistance describes a fabric's resistance to the flow of water vapor. In scientific terms, the water-vapor resistance is defined as the water-vapor pressure difference between the two sides of the fabric divided by the resultant evaporative heat flux per unit fabric area in the direction of the water-vapor pressure gradient.

Recall, permeability describes the ability of a fabric to allow heat, gases, or liquid to pass through it. This means water-vapor permeability (aka breathability) is a fabric characteristic that depends on water-vapor resistance and temperature. Consequently, water-vapor permeability describes the fabric's ability to let water vapor pass through it. Technically, the change of the mass of water vapor, M_v with time, dt, is given by

$$dM_v/dt = P \cdot A \cdot (e_{inside} - e_{outside})/h$$

Where P, A, e_{inside}, $e_{outside}$, and h are the water-vapor permeability, fabric area, water-vapor pressure at the inside and outside of the fabric, and fabric thickness, respectively.

The normalized unit for water-vapor permeability is called the water-vapor transmission rate aka *moisture transmission rate* (MVTR). WVTR refers to the steady-state rate at which water-vapor flows through a fabric at specified conditions of temperature and relative humidity (typically 25°C, 75%, 1 atm). Here steady-state means that the influx is equal to the outflux. The lower WVTR, the smaller the amount of moisture that passes through the fabric in a given time, dt. Generally, WVTR decreases with increasing fabric thickness and air temperature.

Fabrics with low or no permeability for sweat store moisture on their inside when more sweat is produced than sweat evaporates (Fig. 3.10). Low evaporation with accumulation of sweat may also occur when the relative humidity of the ambient air is high. In both cases, the water-vapor resistance is high. Consequently, the fabric absorbs sweat which can cause stickiness and dampness; the fabric may stick to the skin.

We refer to the amount of water absorbed by the fabric as *water absorption*. Technically, we can calculate the water absorption by dividing the difference between the weights of the fabric with and without absorbed water by the weight of the dry fabric. When a fabric absorbs too much liquid, the liquid spreads over the fabric's surface.

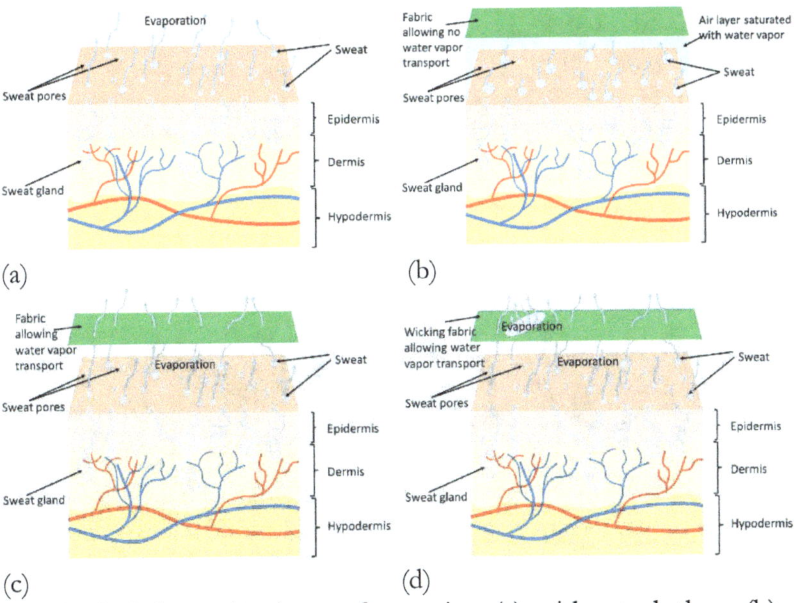

Fig. 3.10. Schematic views of sweating (a) without clothes, (b) with fabric that is non-permeable for water vapor, (c) with fabric being permeable for water vapor, and (d) with moisture-wicking fabric.

As if wet clothes weren't bad enough, water-vapor and moisture permeability also affect the thermal wear comfort. The accumulated moisture namely reduces the thermal insulation. A wet cotton T-shirt has zero insulation.

Two processes, *wetting* and *wicking*, control the liquid moisture flow through textiles. Wetting of textiles refers to the displacement of air in the fabric pores at the fabric-air interfaces with a liquid (Fig. 3.11). As a result, fabric-liquid interfaces form. The surface energies of the fabric and liquid control this initial process of fluid spreading. In other words, wettability is the potential of a fabric to interact with water or sweat.

Wicking refers to the spontaneous flow of moisture (water, sweat) in textiles due to *capillary forces*. The capillary forces drive the liquid into the capillary spaces (pores). Wicking occurs at multi-scales. At the *yarn scale*, moisture wets the fibers and creates a thin film. Due to the presence of nearby fibers, *menisci* form where capillary forces drive the further moisture uptake of the yarn (Fig. 3.11). These processes depend on the yarn's wettability, twisting level, and hygroscopic properties. At the *fabric scale*, the inter-yarn geometry like fabric structure (e.g., weave, knit), direction and pick density define the overall wicking behavior. The smaller the gaps between individual fibers, the larger the capillary forces become. As a result, moisture transport is best for fine fibers. Furthermore, the spread may occur at different speeds in warp and weft directions [99]. This difference can be caused by different materials, yarn count in warp and weft direction, and knit/weave structure. Note that the water-transmission rate of a fully-wetted fabric is identical to that of a free liquid surface.

Fig. 3.11. Schematic view of a cross-section through a woven fabric showing the menisci (blue) formed at the inter-yarn contacts between warp (black thick up- and down line) and weft yarns (dots). Here it is assumed that the water flows along the weft yarns faster (therefore surrounded by blue) than along the warp yarns.

High wicking helps to transfer perspiration through fabric quickly. Typically, as you start sweating, wicking is slow for fabrics consisting of *hygroscopic* fibers because it takes time to absorb the water. Consequently, the slow rise of water in cotton fabric, for instance, is due to moisture absorption and swelling of fibers. The total *absorbency* of a fabric depends on the air proportion within the fabric. It measures of fabric's capacity to hold sweat.

Recall that the bulk porosity of woven fabrics depends notably on the weave parameters (Fig. 3.4b). Two elements in a weave, the float and interlacing points, can increase the water transfer to a certain degree. Studies with woven cotton fabric in different weave patterns, for instance, revealed that fabric with more floats has a high water-transfer speed. The wicking rate increases with an increase in float because the increased length-floating-point fraction creates a loosened structure with more macropores. Consequently, woven fabric with more float has nonuniform capillary channels and more air space for evaporation. As a result, air convection between the skin and the inside of the fabric can increase the transfer to the outside of the fabric. The enhanced convection supports the take-up of water vapor by ambient air, which accelerates evaporation. This means float can enhance the total water transfer in the wicking–and–evaporating process more than the interlacing points [48]. It also means that fabric with evenly distributed floats has slow wicking rates.

Fabrics can regain moisture. *Moisture regain* refers to the amount of moisture that dry fiber would absorb from the air at 70F (21.1°C) and 65% relative humidity. It is expressed in percent of the dry fiber weight. The higher the moisture regain, the longer the fabric feels dry.

Water absorbency, WA, refers to the liquid absorptive capacity of a fiber. It is expressed as the difference between the weight of the wet, m_{wet}, and dry fabric, m_{dry}, as a percentage of the dry fabric weight

$$WA = (m_{wet} - m_{dry}) \cdot 100 / m_{dry}.$$

Water absorbency describes the combined effect of water absorption by fiber and filling up the inter-fiber and inter-yarn pores. In simple words, it gives the water amount that dry fiber retains after soaking in water and drip drying.

3.2.2 Role of Fiber Type and Fabric Structure for Moisture Transport

Typically, in knits, air permeability and relative water-vapor permeability increase with increasing stitch length. Therefore, water-vapor permeability is higher for loose than tight knits. This means the structural properties influence moisture transport. In the case of woven fabrics, these properties are yarn and fiber type, yarn count, number of yarn folds, weft density, weave patterns as well as fabric treatment and blend ratio, if applicable.

Water-vapor resistance increases with increasing air entrapment, fabric thickness, and weight for all fabrics consisting of just one type of fiber. However, the increases depend on the fiber type (Fig. 3.12a). Furthermore, water-vapor resistance is independent of fabric structure (Fig. 3.12b). Water-vapor resistance is low for Tencel and viscose because of their high moisture regain. The same is true for *hydrophilic* polyester (red dot at 0.6 mm, 2 m^{2}Pa/W) and wool. Furthermore, water-vapor resistance is lower for fabrics made of fine than thick yarns [41].

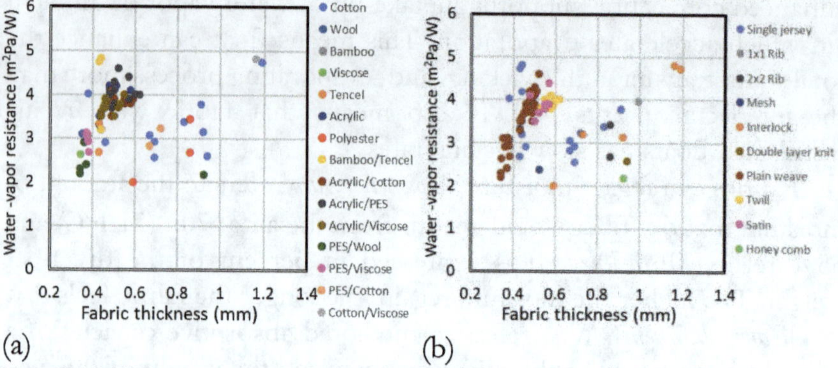

(a)   (b)

Fig. 3.12. Water-vapor resistance vs. fabric thickness for various fabric (a) raw materials, and (b) structures. PES is polyester. Modified after [82]. Data from [26], [29], [33], [41], [76], [100], [101].

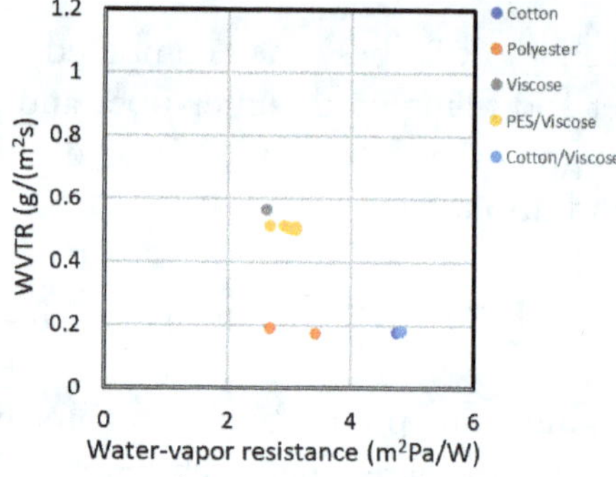

Fig. 3.13. Water-vapor resistance vs. WVTR for various fiber types. Data from [26], [102].

Unfortunately, only a few studies reported both water-vapor resistance and water-vapor permeability/WVTR/relative moisture transfer. Nevertheless, we see that water-vapor resistance increases when WVTR decreases (Fig. 3.13). Consequently, you feel more uncomfortable in heavy, thick polyester and cotton clothes than the thinner polyester/viscose clothes (cp. Figs. 3.12a, 3.13). A high water-vapor resistance namely makes it difficult to transport evaporated sweat away from your skin. Therefore, fabrics used in sportswear or summer clothing require low water-vapor resistance. The increased WVTR of fabrics with low water-vapor resistance namely results in high breathability. Consequently, during high body activity and/or increased sweating, an increased water-vapor permeability results in improved wear comfort because the sweat can vanish fast from your skin.

Because of its fine yarns, twill has a higher water-vapor resistance than basket- or break-twill weave. Finer yarns, namely, have a comparatively large surface area to resist water vapor and thermal energy. However, when considering thickness and porosity, the three different weaves yield similar water-vapor resistance [40].

Now let's look at fabrics close in fabric thickness (Fig. 3.14) to assess the impact of yarn raw material and fabric structure on WVTR. At the same thickness, WVTR can vary more than a factor of 100 among different fabrics (Fig. 3.14a, b). WVTR decreases non-linearly with fabric thickness. The degree of decrease depends on the fiber type.

As expected from the results for water-vapor resistances, fabric structure influences WVTR marginally when compared to fabric thickness or fiber type. At similar fabric thickness, plain weaves, for instance, show low to high WVTR depending on the fiber type.

However, the WVTR of cotton or polyester single-jersey is higher than for 1×1 rib and interlock fabrics of the same material [70].

Polyester fabrics with rainfall knit or waffle knit have higher water-vapor resistance than single-jersey, weft knit, warp knit raschel, and interlock knit fabrics [24]. The reason is the higher thickness and weight of the former than the latter fabrics.

At the same fabric thickness, cotton fabric has much lower WVTR than cotton/bamboo fabrics. This means blending cotton with bamboo can increase the WVTR. Silk/wool blends have the highest WVTR of the fabrics at the same fabric thickness (Fig. 3.14a).

In the case of blends with highly hydrophilic fibers, there is a direct relationship between the blend ratio and WVTR. For example, the water-vapor resistance of polyester/viscose blends decreases with increasing percentage of viscose (Fig. 3.15) because viscose is a highly hydrophilic fiber. Furthermore, the fabric thickness of polyester/viscose blends increases slightly with increasing viscose content. However, a 100% viscose fabric made the same way is thinner than the polyester/viscose blends (Fig. 3.15).

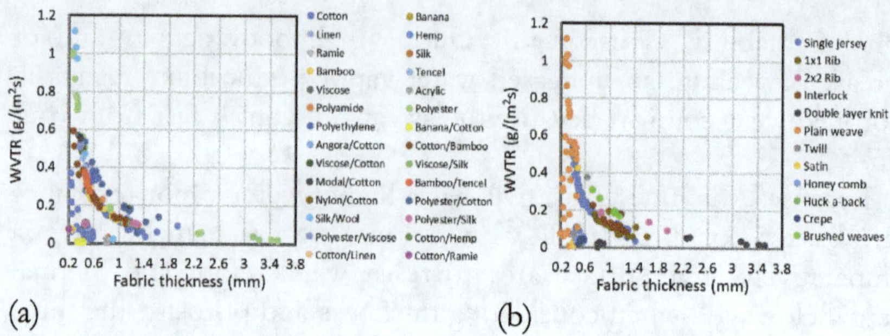

Fig. 3.14. Water-vapor transmission rate (WVTR) vs. (a) fabric thickness for various fiber materials and blends, and (b) fabric thickness of various fabric structures. Modified after [82]. Data from [14], [15], [25], [29], [32], [34], [36], [39], [41], [42], [54], [73], [75], [77], [78], [86], [89], [90], [95], [96], [97], [98], [102], [103].

Banana fiber has high air permeability and is very hydrophilic. Therefore, a 50/50 banana/cotton satin has the highest WVTR. Whilst 25/75 banana/cotton plain weave has the lowest air and water-vapor permeability [98], and hence WVTR. Consequently, adding a hydrophilic fiber to a fiber of choice, which is less hydrophilic, can improve the breathability of the end product (Fig. 3.14).

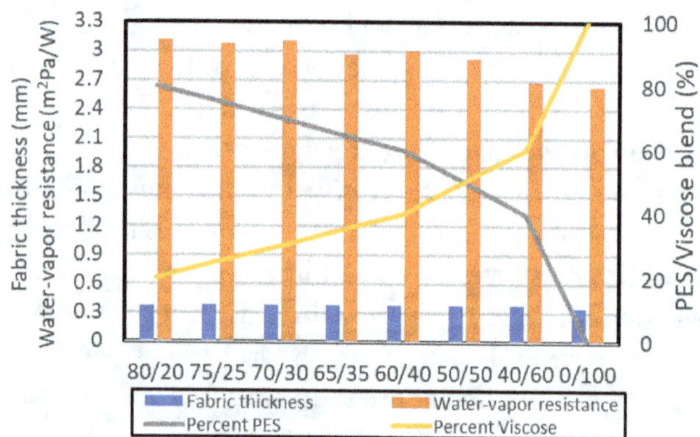

Fig. 3.15. Fabric thickness and water-vapor resistance of polyester/viscose blend with various blend ratios and comparison with a 100% viscose fabric. All fabrics were produced the same way except for the ratio of fibers. From [82]. Data from [102].

Hygroscopic fabrics absorb water vapor from the humid air close to the sweating skin and release it in comparatively drier air. Consequently, when the ambient air is drier than the air inside the clothes, the transfer of water vapor to the ambient air increases as compared to non-absorbing fabrics. Therefore, hygroscopic fabrics decrease moisture built-up. The thinner a fabric, the greater its absorption rate. Absorption rates depend on the blend ratios.

On the contrary, fabrics with low or no *hygroscopicity* have high water-vapor resistance. Generally, water entrapment in pores increases with increasing porosity.

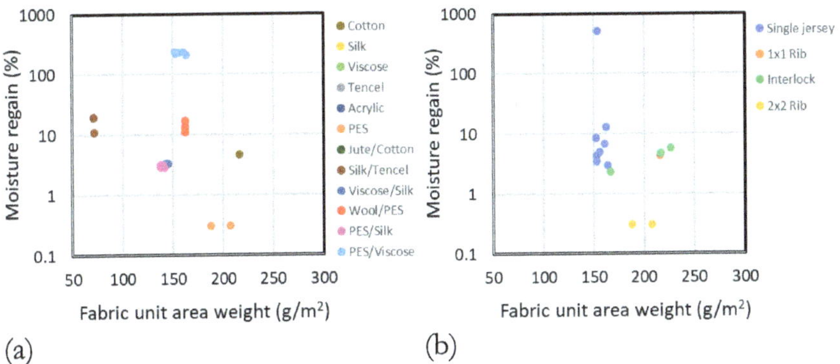

Fig. 3.16. Moisture regains for various fabric (a) fiber types, and (b) structures. PES is polyester. Data from [26], [102].

Silk fiber has a high density of hydrophilic groups and nanopores. Therefore, silk fabric can contain between 10% and 30% of its weight in moisture, but about 11% is typical [104]. Viscose fiber has higher moisture regain than silk (Fig. 3.16). Moisture regain of all silk and silk/viscose blended fabric exceeds that of cotton fabric while the moisture regain of silk/polyester blended fabrics is proportional to the hygroscopic fiber (silk) content.

Polyester fails to absorb perspiration because it lacks moisture-retention capability [105]. Moisture regain and moisture absorbency of polyester/viscose fabrics increase linearly with increasing viscose content because viscose is hydrophilic. However, once the water molecules are in the capillaries, they form bonds with the absorbing fiber molecules due to the high *water affinity* of viscose. This process inhibits the capillary flow along the pores. Consequently, water absorption governs the movement of water in the fabric. On the contrary, *hydrophobic* polyester creates no bonds with water molecules. Consequently, blending viscose with polyester provides channels for water movement and high wicking.

Wool and felt fabrics can regain huge amounts of moisture without feeling damp. Moreover, wool is hydrophobic. Consequently, these fabrics have been favored for winter clothes for centuries in maritime climates (see Ch. 1). We can classify a 100% wool fabric as waterproof due to its very slow rate of water absorption, low spreading speed, and lack of one-way transport and water penetration [100].

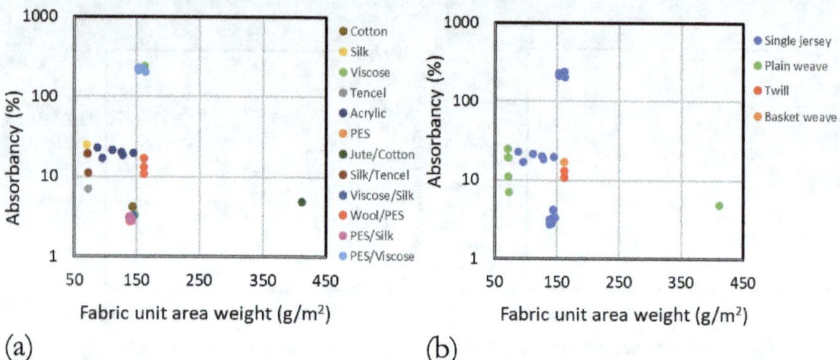

Fig. 3.17. Absorbency of various fabrics sorted by (a) fibers, and blends, and (b) structures. Y-axis is logarithmic. Data from [28], [40], [77], [86], [102], [106].

Absorbency differs more than a factor of 100 among various fabrics depending on the fiber, blend, and blend ratios (Fig. 3.17). In the case of cotton/bamboo blends, for instance, absorption rates increase with bamboo content [86].

Typically, wicking is given for lab conditions (25°C, 1 atm, 65% relative humidity). However, capillary pressure and water-vapor permeability which govern wicking behavior, both depend on the saturation within the fabric. Water-vapor permeability namely decreases when saturation decreases. Once saturation falls below a critical value, yarns with smaller effective capillary radii sustain higher water-vapor permeabilities than those with larger effective capillary radii. In simple words, at ambient temperature and relative humidity conditions, the wicking behavior of a fabric might differ from that determined in the lab. Unfortunately, no guidance exists on which fabric is best for which environmental conditions. Furthermore, washing alters the wicking behavior too (Fig. 3.18c).

Wicking rates also depend notably on fiber type, blend ratios and fabric unit area weight (Fig. 3.18a, b). Polyester/spandex blends, for instance, have very low wicking rates. Wicking increases for viscose/silk blends strongly with an increase in viscose content. Blending cotton with polyester increases wicking with an increase in polyester. Studies on polyester weft-knitted fabrics for sports T-shirts showed that increasing fabric weight increases air resistance; a higher porosity means faster drying; and shorter water-absorption time means a faster vertical wicking rate [107]. Cotton fabric has the highest sweat-holding capacity followed by silk and polyester knitted fabric. Silk fabric has good wicking, WVTR, and total absorbency.

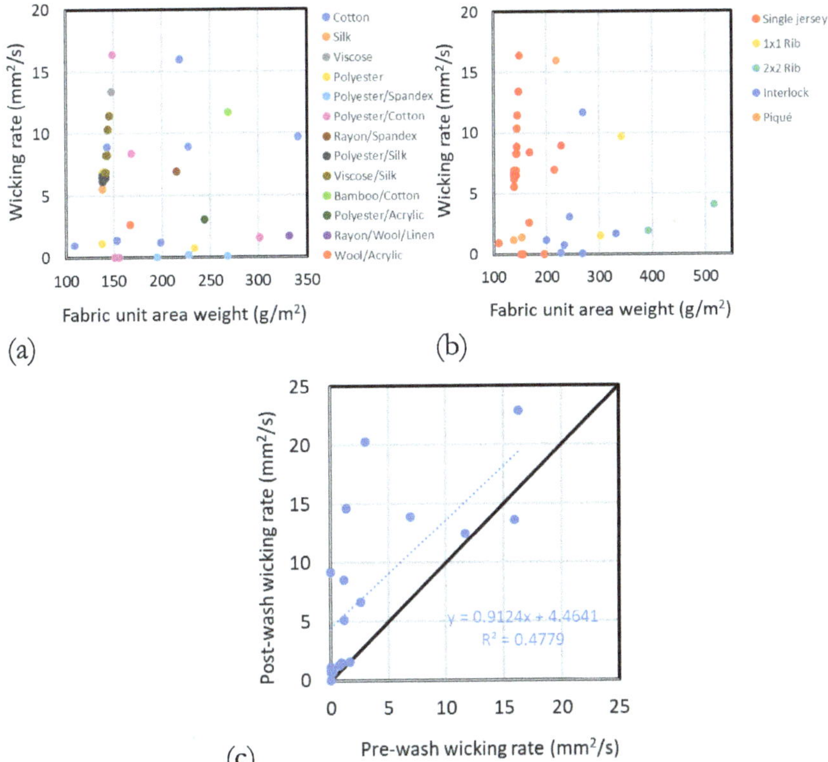

Fig. 3.18. Wicking of fabrics sorted by (a) fiber types and blends, (b) structures, and (c) wicking rates before vs. after washing. Fiber blends differ in ratios. Data from [95], [108].

3.2.3 Moisture Management Indexes

Fabrics with high moisture absorption permit the quick release of perspiration from the skin to keep the wearer dry. Low wetting times indicate fast absorption of sweat. Wetting times change with the yarn count of knit fabrics. The spreading speed of the fabric affects moisture comfort. The maximum wetting radius of a fabric influences the spreading area of sweat on the fabric surface. Consequently, as the maximum wetted radius increases, the drying time of the fabric decreases. As a result, a great fabric for high activity and/or strong sweating in summer requires good air permeability, low thermal- and water-vapor resistances as well as good *moisture-management* properties to release excess of heat, and keep the wearer dry.

A series of *moisture-management indexes* exists for describing different aspects of the fabrics' moisture-management properties. They are assigned grades (Tab. 3.2) ranging from 1 (very poor) to 5 (excellent). These indexes are the wetting time (WT_T, WT_B), moisture-absorption rate (AR_T, AR_B), maximum wetted radius (MWR_T, MWR_B), moisture-spreading speed (SS_T, SS_B), and accumulative one-way liquid transport capacity (OWTC). Here the subscripts T and B refer to the top- and bottom sides of the fabric. The *overall moisture-management capacity* (OMMC) of a fabric is given by

$$OMMC = 0.25 \cdot AR_B + 0.5 \cdot OWTC + 0.25 \cdot SS_B.$$

Fiber type, tightness, and blend ratio govern the OMMC. At about 50 g/m^2 fabric area weight, for instance, the OMMC of plain weave wool, cashmere/wool, and cashmere fabrics ranges from below 0.4 to slightly above 0.6 (Fig. 3.19). Polyester/cotton fabrics have better OMMC than polyester/bamboo fabrics because bamboo absorbs more water than it spreads it. While both cotton and bamboo fibers are both hygroscopic, their moisture regain is 8.5% and 13.5%, respectively. Increasing the bamboo content of polyester/bamboo blends yields OMMC similar to polyester/cotton blends. This means we can achieve increased sweat-evaporation rates, which enhances the wear comfort in warm environments. Therefore, polyester/bamboo blends could replace polyester/cotton blends for apparel because of their better strength, low stiffness, low thermal resistance, and similar moisture management properties [62].

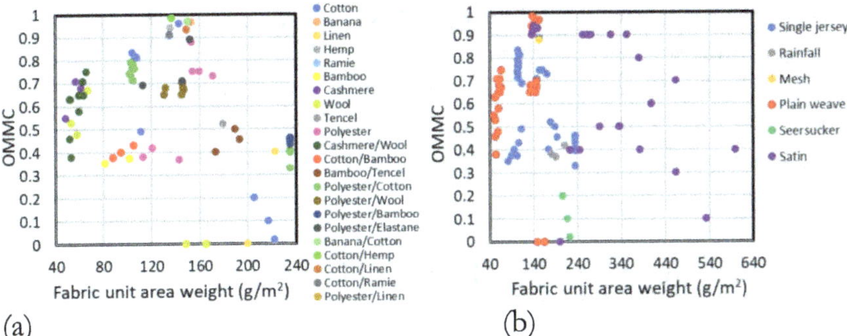

Fig. 3.19. OMMC for various (a) fiber types and blends, and (b) fabric structures. Data from [24], [26], [29], [62], [90], [109], [110], [111], [156].

Table 3.2. Grading scales for moisture-management properties. WT, AR, MWR, and SS are the wetting time, absorption rate, maximum wetted radius, and spreading speed. After [105]. The letters VP, P, G, VG, E stand for very poor, poor, good, very good, and excellent. NW, VS, S, M, F, VF indicate no wetting, very slow, slow, medium, fast, very fast, large, and very large.

			Grade		
Index	1	2	3	4	5
OWTC (%)	< -50	-50–99	100–199	200–400	> 400
	VP	P	G	VG	E
OMMC	0–0.19	0.2–0.39	0.4–0.59	0.6–0.8	>0.8
	VP	P	G	VG	E
WT (s)	≥ 120	20-119	5-19	3-5	<3
	NW	S	M	F	VF
AR (%/s)	0-10	10-30	30-50	50-100	>100
	VS	S	M	F	VF
MWR (mm)	0-7	7-12	12-17	17-22	>22
	NW	S	M	L	VL
SS (mm/s)	0-1	1-2	2-3	3-4	>4
	VS	S	M	F	VF

Of all fabrics, cotton seersucker and heavy wool fabric have the worst moisture management (Fig. 3.19). Nevertheless, seersucker is a favorite in hot humid climates due to its thermal comfort. Wool fabric with low density for its thickness is basically waterproof and a favorite in wet, humid weather.

In the case of denim, the types of weft yarns and washing treatments influence the moisture-management capacity [99]. For woven hemp fabrics, the OMMC increases as the yarn number increase [112].

3.3 Conclusions on the Overall Thermo-Physiological Comfort of Fabrics

Dressing appropriately for various weather conditions requires a clear understanding of the heat and moisture transfer through the various types of fabrics. The temperature difference between the skin surface and environment dominates conduction, convection, and radiation, and, hence, dry heat transfer. On the contrary, the water-vapor pressure difference between the skin surface and the environment relates to the latent heat and moisture transfer.

To judge the combined heat and moisture transfer behavior we can calculate a *water-vapor permeability index*, i_m, (aka *moisture permeability index*) from the thermal resistance, R_{ct}, and water-vapor resistance, R_{et}, as [113]

$$i_m = 60 \cdot R_{ct}/R_{et}.$$

This (dimensionless) index ranges from 0 (impermeable) to 1 (permeable) and indicates the degree of fabric permeability. The water-vapor permeability index values (Fig. 3.20) depend on fabric structure (e.g., fabric weight and thickness, porosity, fiber density). Did you know that ironing increases thermal conductivity, thermal absorptivity, and water-vapor permeability, while it reduces the thermal resistance [15]?

What are the consequences of a fabric's heat- and moisture transfer for the wearer? In winter for protection from the cold, you should prefer double-jersey structures, 1×1 rib, or interlock fabrics due to their high thermal insulation. At first sensation, 1×1 rib fabrics have a warmer feeling than an interlock-knit fabric. Blends with low cotton fraction are convenient for winter clothes. Garments from wool blends with just a small fraction of cashmere keep you warmer longer than garments with the same fabric structure made from pure wool.

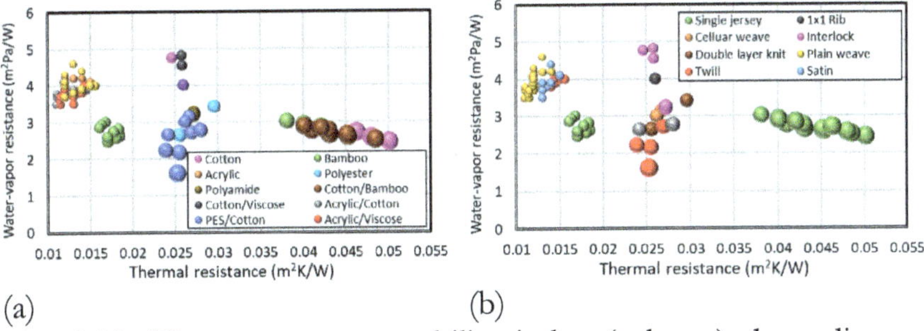

Fig. 3.20. Water-vapor permeability index (spheres) depending on thermal and water-vapor resistance for various fabric (a) types and (b) structures. The largest and smallest spheres indicate 1 and 0.167. Fabrics of the same material and structure differ by weave or knit density. Data from [10], [26], [32], [33], [62], [86].

For sport and summer clothing, single-jersey structures should be your first choice because of their better moisture-management properties than other knits. Polyester/cotton blends with high cotton fractions are suitable for summer clothes as well. Polyester and cotton/Coolmax fabrics with float structures have good liquid moisture-transfer properties as well, as high capillary-absorption characteristics [114].

Of course, yarn twist, weave or knit density, fabric thickness, etc. affect the transfer of heat, water vapor and moisture through fabrics. Based on just the materials' resistance to dry heat and latent heat transfer, we find the following: While the water-vapor resistance of single-jersey fabric ranges between 2.5 and 3 m²Pa/W for the fabrics in Figure 3.20, its thermal resistance ranges from 0.016 to 0.05 m²K/W with decreasing tightness of the knit. Consequently, tight single-jersey cotton, bamboo, and cotton/bamboo fabrics are better for summer, while loose versions of these fabrics are better for winter. Acrylic, acrylic/polyester, and acrylic/viscose plain, twill, and satin fabrics have high water-vapor resistance, but low thermal resistance. Wool keeps you warmer than cotton and polyester, with wool from sheep who lived in cold climate regions holding you warm longer than wool from sheep in warm climate regions.

Fabrics from regenerated cellulose (rayon, bamboo, Tencel, Modal) are soft, breathable (high water-vapor permeability), very easy to dye, and moisture-wicking. Furthermore, they are anti-allergic due to their inherent anti-bacterial property. Bamboo rayon is silkier and smoother than rayon and feels cool to the touch. Therefore, bamboo rayon is a favorite in the bedding industry for summer sheets. Tencel is non-clinging.

Looking at the thermal and water-vapor resistances, as well as air and water-vapor permeability, Tencel single-jersey, and polyester mesh-knit fabric or their combinations are preferable for sports gear, and under weather conditions yielding to high perspiration [24]. In the case of Tencel/bamboo single-jersey fabrics, WVTR increases with increasing Tencel content due to the decrease in fabric weight and thickness [29]. Therefore, Tencel/bamboo blends are also great for situations when you anticipate strong sweating.

The thickness and weight of polyester/bamboo knit fabrics decrease with increasing bamboo content. Air permeability, moisture absorption, and moisture-management capability of polyester/bamboo and polyester/cotton blended fabrics increase, while their thermal resistance decreases with increasing bamboo and cotton content, respectively. As the polyester content decreases, polyester/cotton fabrics have higher thermal resistance than polyester/bamboo fabrics because of the higher air permeability of polyester/cotton blends. Consequently, the thermal resistance of polyester/cotton fabrics increases because of the increase in entrapped air and fabric thickness [62]. Because of their moisture-management properties, polyester/bamboo and polyester/cotton fabrics are excellent for comfortable summer apparel.

Despite cotton has a high affinity to moisture, the blend ratio of polyester/cotton fabric insignificantly influences the water-vapor resistance of the fabrics [10]. As a result, 65/35 polyester/cotton fabrics are convenient for winter clothing, while 33/67 polyester/cotton yarns are suitable for summer clothing. The latter is due to cotton acting as microclimate by means of its high WVTR, especially, at high temperatures.

Cotton-in-warp and bamboo-in-weft plain woven fabric, for instance, has higher air permeability and wicking rates, and lower thermal resistance than plain 100% bamboo or 100% cotton weaves. Pure cotton has the lowest wicking rate of the three [31], [29].

Use of bamboo-viscose, soybean or Seacell for the weft, and cotton for the warp marginally reduces thermal resistance and increases thermal conductivity as compared to 100% cotton [89]. According to the thermal absorptivity values, 50/50 bamboo/cotton or 50/50 soybean/cotton fabrics provide a cooler sensation than 100% cotton fabrics. Therefore, these fabric blends are very suitable for warm conditions, and bedding in hot climates or summers. Out of these three blends, cotton/soybean[32] fabric has the highest breathability, and smoothest surface [89].

Recently, the textile industry often replaces cotton with flax/cotton or cotton/hemp blends in the production of denim fabrics. Blended denim has superior moisture-comfort properties (higher water absorbency along with quick-drying) and thermal resistance than cotton denim, while having similar tactile comfort [115]. Blending cotton with flax or hemp also improves the softness of fabrics with 2/1 twill and 3/1 twill weaves.

Typically, fabrics with high bamboo content have notably higher air and water-vapor permeability, water absorbency, and wicking, while lower thermal resistance than cotton fabrics produced the same way [31].

Low weft density in 2/2 twill and matt twill fabrics results in smoother surfaces, and lower water-vapor resistance than cellular or diced weaves [10]. The smoother surface means a cool sensation on the first touch. For these reasons, the low weft density 2/2 twill and matt twill fabrics are great for summer clothing.

The puckered texture of seersucker creates air spaces between the skin and fabric. These spaces promote air circulation and avoid adherence of the whole fabric to the skin. As a result, the wearer stays cool in hot weather. The repeat of puckered strips reduces the liquid moisture-transfer properties, in particular, the absorption rates, maximum wetted radius of the top and bottom fabric surfaces, the spreading speed on the top surface, and the moisture transport from the inner to the outer side of cotton seersucker. The OMMC obtained for different seersucker weaves is poor to very poor ranging between 0.02 and 0.22 [110]. Washing can improve the moisture-transport properties of seersucker fabric.

3.4 Why Fabric Treatments?

In the textile industry, many fibers or yarns are treated to change and/or improve their properties. Liquid ammonia treatment, for

---

[32] Given that still today people starve from hunger, it might be questionable whether using soybean cake for fiber production is an ethical choice.

instance, alters the crystal structure of hemp fibers. This treatment turns some of the cellulose into cellulose III, which decreases the crystallinity from 66.1% to 57.4%. As a result, hemp-woven fabrics show improved liquid moisture-management properties. Treating polyester-fiber fabrics with hydrophilic finishes improves sweat distribution.

Industrial washing treatments (e.g., rinse, stone, bleach washes) improve the comfort properties of denim fabrics in a similar way as blending cotton with hemp or flax [116].

The so-called *sol-gel technique* serves for conferring new functional properties to fabrics, such as dye fastness, antiwrinkle finishing, hydrophobicity, antimicrobial properties, or UV-radiation protection [94]. Because sol-gel coatings have a porous structure, these finishings can alter the thermal insulating properties, in particular, those related to thermal radiation, i.e., body heat, which is in the IR range. Treatment with *energy reflection chemicals* (ERC) like acetic acid serves to reduce heat from this IR-radiation (Ch. 2). While, ERC has no notable impact on air permeability, water-vapor resistance [10], and OMMC [100], it can affect thermal properties.

Highly reflective metallic fibers as inside fabric layer can reduce body radiative heat loss because they reflect the body heat back to the skin. As a result, the wearer feels warm and comfortable in cold weather longer than without such treatment.

3.5 Increasing Insulation by Layering

Whenever there is a temperature contrast, there occurs a *net heat flow* from the warmer to the colder object to achieve equilibrium between the objects of different temperatures; the same is true for moisture fluxes going from the moister to the drier object. Applied to clothing these objects are the body and clothing and the clothing and air.

Recall, thin air layers are bad conductors for heat, i.e., they are great thermal insulators. Consequently, sleek layering can increase thermal insulation, and thermal comfort in cold weather, and looks incredibly stylish. However, when layering clothes, the overall conductivity through the layers depends on the air space between layers, the fabric materials, and structure.

Unfortunately, air is very mobile. Therefore, for protection from cold weather, you have to establish layers of "still air" by the use of clothes. In windy, cold climate, in addition, you have to stop the wind from penetrating through and/or entering the clothing.

In summer - and during high-energetic activity in winter -, removal of sweat is important. Therefore, using several layers of clothes with different properties may still reduce the heat loss, but the body's internal heat production may not match the heat loss. The degree of insulation depends strongly on the thickness and fit of the clothes, the amount of air trapped between layers or within the fibers of the clothing, and the water-vapor permeability of the outfit.

In cold climate, low water-vapor permeability and sweating can result in wetting of clothes, which reduces their thermal insulation proportional to the moisture retention. Therefore, breathability of your clothes is very important. When you run at 30 below (-34.4°C) to catch the bus, you start sweating. In the warm bus, sweating continues. When your clothing is not semi-permeable to water-vapor transfer from the inside to the outside, you have the recipe for catching a cold. Once you are again outside, the water-vapor from evaporated sweat may condense, thereby damping your clothes. Eventually, the water may even freeze. Therefore, look for semi-permeable fabric that keeps outside moisture out, but permits moisture from sweating to leave.

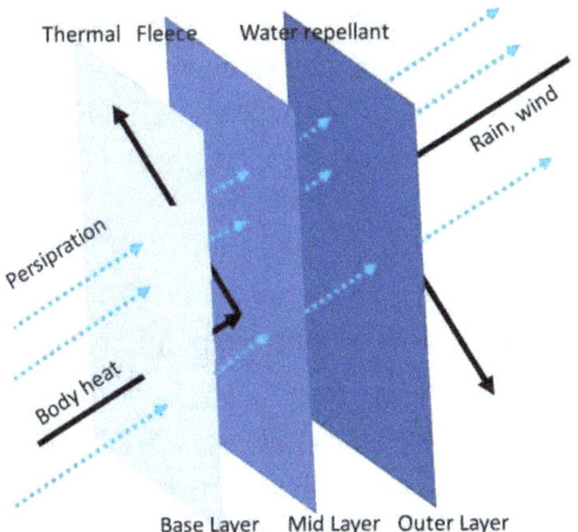

Fig. 3.21. Schematic view of layering for humid, windy weather with temperatures above 32F (0°C).

Both thermal insulation and water-vapor permeability increase and then decrease with increasing thickness of the air layers between clothes. Consequently, there is an optimum for the still air-layer thickness. This knowledge translates as follows: Your clothes must have the right size. Too tight clothes can lose more than half of their thermal insulation capability due to the squeeze. Squeezing inhibits the creation of an insulating air layer (Fig. 3.21) between your skin and the fabric, and/or between different fabric layers. Moreover, (too) tight clothes build cold bridges.

In calm weather, the highest thermal insulation occurs for an air-layer thickness of 0.39 inch (1 cm). In windy weather, the optimum thickness is about 0.24 inch (0.6 cm). I will explain the reason for these differences later (Ch. 5). You can achieve the required air-layer thickness by three to five layers of clothes. The outer layers shall not compress the layers worn underneath. Only underwear should be snug to your skin – a bra or tummy control would lose all their purpose if they were loose, right? Ideally, the mid-layer reflects body heat back to the base layer (Fig. 3.21). The outer layer should be wind- and rain-proof.

If you hold the thought, you already get the idea that layers mean adding additional thin insulation. Similar to the dry (aka *sensible*) heat transfer through clothing, the evaporative heat transfer seems to depend linearly on the insulation of the outfit.

The fabric closest to the body has to be moisture-wicking to avoid that the fabric gets wet (Fig. 3.21). Modal, rayon, viscose, Tencel, cotton, and bamboo take up sweat. However, once they are wet, their insulation is zero! Consequently, these materials are a bad choice for thermal comfort at temperatures below 50F (10°C) or so. On the contrary, merino wool, for instance, can take up a lot of moisture without feeling wet.

3.6 Climate Rooms and Sweating Manikins - Fit Matters Not Only for Style

Researchers use *controlled-climate chambers*[33] in combination with so-called *sweating-manikin systems* to assess entire outfits or components of outfits for heat and moisture management, insulation, and breathability. These climate-controlled rooms, namely, permit them to measure under exactly repeatable climatic conditions, while changing different aspects of clothing. The manikins have human size and form

---

[33] Climate chambers are rooms wherein prescribed temperature and moisture conditions can be created.

and can move to simulate activities like walking, sitting, working at a desk, etc. They are heated to skin temperature. To simulate sweating a pump supplies preheated water from a reservoir outside of the climate-controlled room to the "sweat glands" on the surface of the manikin. A custom-made bodysuit serves as "skin". It consists of wicking material to distribute the moisture evenly over the entire manikin. Each of the sweat glands is individually calibrated so the researcher can set the desired sweat rate via the control software. The software also records the heat loss of the manikin.

Obviously, these sweating-manikin systems permit much better data of human heat loss than measurements performed on the fabric alone. Furthermore, the researcher also can examine influences of fit, clothing construction, design, trapped air layers, motion, ventilation, layering, etc. due to the possibility to isolate these effects [117].

Studies can also assess the impact of so-called *cold gaps* or *cold bridges* on body-heat loss. The term "cold gaps" refers to regions of increased heat loss (lower insulation) than elsewhere. Examples of cold gaps exist at the hems at the legs of pants, sleeves, as well as the neck and hems of jackets/coats. Any motion of the wearer permits cold, dry air to enter the clothes at the hems; i.e., motion acts like ventilation, and mixes the cold air with the warm air inside the clothing. Rib-knit cuffs inside of sleeves, legs, and even inside jackets, and parkas just above the hem can mitigate the impact of cold gaps.

4 METHODS TO SELF-ASSESS THE THERMO-PHYSIOLOGICAL COMFORT OF YOUR OUTFIT

4.1 Measuring Clothing Insulation and Heat Transfer

The textile branch measures clothing insulation in clo (1 clo=0.155 $K \cdot m^2/W$). The clo-unit describes the *sensible* (aka "dry") *heat transfer* by radiation and convection. Recall, this heat transfer takes place between your skin and the ambient environment. A clo is defined as the thermal insulation by clothes that allows a person at rest to stay in thermally comfortable condition in an environment at 70F (21°C) in a normally ventilated room (0.22 mph, 0.1 m/s air movement) with a relative humidity of 50%. At lower temperatures, a person dressed this way feels cold, while at higher temperatures, they start sweating. Typically, the insulation of fabric with 1 clo approximately compensates for a drop in temperature of about 15F (9°C).

Per definition, 1 clo-unit of clothing and air insulation allows 5.55 $kcal/(m^2 \cdot h)$ of heat exchange by convection, Q_{conv}, and radiation, R, for each degree Celsius of temperature difference between the mean skin temperature, T_{skin}, and the adjusted ambient (bulb) temperature ($T_{bulb}=0.5$ ($T_{air}+T_{rad}$). Here T_{air} and T_{rad} are the air temperature and mean radiation temperature, respectively. This definition permits us to estimate the radiative plus convective heat transfer for an average body-surface area of 19.4 square feet (1.8 m^2) as

$$(Q_{conv}+R)=10/clo \ (T_{skin}-T_{bulb}).$$

This means that a still air layer of 0.8 clo by itself restricts the convective and radiative heat exchange of a naked person to about 12.5 kcal/h per °C of difference between the skin- and air temperature. Under the above defined conditions, you can assume the following values for basic insulation:

Naked body	0 clo
Summer clothes	0.6 clo
Ski clothes	2 clo
Light polar clothes	3 clo
Heavy polar clothes	4 clo
Polar down duvet	8 clo

The *ASHRAE-55 2010* standard of the American Society of Heating, Refrigeration, and Air-Conditioning Engineers (ASHRAE) points out three methods for estimating the insulation by clothes [118]. When your clothing ensemble matches acceptably well with one of those described in Table 4.1, you can estimate its potential insulation by adding the respective values. Various studies namely suggest that the insulating effect of clothing increases linearly with the thickness of the clothing worn. Consequently, you can add or subtract the insulation effects from garments in Table 4.2 from the outfits listed in Table 4.1. You can also assess the insulation of your outfit as a combination of the individual outfit pieces by using the values listed in Table 4.2.

It is hard to generalize the above findings to very thin layers of fabric like underwear. A typical clothing insulation is about 4 clo per in (1.57 clo per cm) of insulation-layer thickness. Even a thin fabric may allow for a still air layer of optimal thickness (~0.24 in, 0.6 cm). While underwear itself contributes little to the intrinsic insulation (Tab. 4.2), it can create an insulating still air layer. Therefore, it is not only from a hygienic point-of-view a *Must* to wear underwear.

Typically, the total intrinsic insulation is approximated to be 80% of the sum of the individual items. This approximation considers an average loss of 20% of the insulation due to the compression of one layer by the next. In reality, the actual reduction depends, among other things, on the type of fibers, fabric weight and structure, fit and cut of the attire, and the use of non-fibrous layers.

According to the ASHRAE-55 standard, it is only acceptable to apply a single, representative mean value for everybody when a person can freely adjust their clothes to suit their thermal preferences. However, when living in cold climate, for instance, your clothing must be adaptable and adjustable to indoor and outdoor conditions to achieve thermal comfort. At the same time, at least your indoor outfit must meet the work dress code and/or social expectations for the occasion [119]. On top of this, you want to be fashionable and stylish. For your quality of life, meeting your personal style preferences is also a *Must*.

According to surveys, most people associate thermal comfort with temperatures between 60 and 81F (15-28°C) which would be a comfort range of 21F (13°C) [118]. Most people feel comfortable within 68 and 80.2F (20-27°C) at relative humidity between 35 and

60% [120]. However, the physiological comfort range of temperatures, within which the human body achieves successful temperature regulation without shivering, or uncomfortably cool fingers, toes, hands, and feet, and without sweating, is much smaller than the surveys suggest. According to ASHRAE, this comfort range is as small as 6F (3.3°C). Therefore, ASHRAE specified that air-conditioned space should have temperatures between 72 and 78F (22.2-25.5°C) for people wearing the typical professional white-collar office outfit of a long-sleeved shirt with pants/skirt. Such an outfit corresponds to 0.6 clo of intrinsic insulation (cf. Tabs. 4.1, 4.2).

Table 4.1. Typical insulation obtained by various outfits. Modified after [118].

Outfit description	Insulation (clo)
Walking shorts, short-sleeved shirt	0.36
Trousers, short-sleeved shirt	0.57
Same as above, plus suit jacket	0.96
Trousers, long-sleeved shirt	0.61
Same as above, plus vest and T-shirt	0.96
Trousers, long-sleeved shirt, long-sleeved sweater, T-shirt	1.01
Same as above, plus suit jacket and long underwear bottoms	1.30
Sweat pants, sweat shirt	0.74
Long-sleeved pajama top, long pajama trousers, short 3/4 sleeved robe, slippers (no socks)	0.96
Knee-length skirt, short-sleeved shirt, pantyhose, sandals	0.54
Knee-length skirt, long-sleeved shirt, full slip, pantyhose	0.67
Knee-length skirt, long-sleeved shirt, half-slip, pantyhose, long-sleeved sweater	1.10
Knee-length skirt, long-sleeved shirt, half-slip, pantyhose, suit jacket	1.04
Ankle-length skirt, long-sleeved shirt, suit jacket, pantyhose	1.10
Long-sleeved coverall, T-shirt	0.72
Overall, long-sleeved shirt, T-shirt	0.89
Insulated coverall, long-sleeved thermal underwear, long underwear bottoms	1.37

Table 4.2. Insulation by various garments. Modified after [118].

Garment	Insulation (clo)	Garment	Insulation (clo)
Underwear		Coverall	0.49
Bra	0.01		
Panties	0.03	**Dresses/Skirts**	
Men's briefs	0.04	Thin skirt	0.14
T-shirt	0.08	Thick skirt	0.23
Half-slip	0.14	Thin sleeveless scoop-neck	0.23
Long underwear top	0.20	Thick sleeveless scoop-neck	0.27
Full slip	0.16	Thin short-sl. shirtdress	0.29
Long underw. Bottom	0.15	Thin long-sl. shirtdress	0.33
		Thick long-sl. shirtdress	0.47
Footwear			
Ankle-length socks	0.02	**Sweaters**	
Pantyhose/stockings	0.02	Thin sleeveless vest	0.13
Sandals/thongs	0.02	Thick sleeveless vest	0.22
Shoes	0.02	Thin long-sleeve	0.25
Quilted, lined slippers	0.03	Thick long-sleeve	0.36
Calf-length socks	0.03		
Thick knee-socks	0.06	**Lined Sport Coats, Vests**	
Boots	0.10	Thin sleeveless vest	0.10
		Thick sleeveless vest	0.17
Shirts & Blouses		Thin single-breasted	0.36
Sleeveless scoop-n.	0.12	Thick single-breasted	0.44
Short-sl. knit shirt	0.17	Thin double-breasted	0.42
Short-sl. dress shirt	0.19	Thick double-breasted	0.48
Long-sl. dress shirt	0.25		
Long-sl. flannel shirt	0.34	**Sleepwear/Robes**	
Long-sl. Sweatshirt	0.34	Thin sleeveless short gown	0.18
		Thin sleeveless long gown	0.20
Trousers/Overalls		Short-sleeve hospital gown	0.31
Short shorts	0.06	Thin short-sleeve short robe	0.34
Walking shorts	0.08	Thin short-sleeve pajamas	0.42
Thin straight pants	0.15	Thick long-sl. long gown	0.46
Thick straight pants	0.24	Thick long-sl. short robe	0.48
Sweatpants	0.28	Thick long-sleeve pajamas	0.57
Overall	0.30	Thick long-sl. long robe	0.69

Each change of 0.18 clo compensates for a 1.8F (1°C) change in air temperature. This means the clothing in which a person would feel comfortable depends on the ambient air temperature.

At 59F (15°C), for instance, a resting person would require about 1.9 clo to feel comfortable, while at 82.4F (28°C), 0.15 clo would be enough. At 32F (0°C), a person waiting outside for the bus would need about 4.6 clo insulation. At temperatures below -22F (-30°C), even when acclimated to cold weather, and when wearing the best extreme cold weather protective clothing, inactive people will not achieve thermal comfort at all. The clothing just provides insulation, and serves for short-term protection.

This fact is just another reason, not to throw fashion out of the window when the *Polar Vortex* moves far south. You just cannot achieve thermal comfort at these temperatures. Period. But at least, you can feel uncomfortable in style [119].

4.2 Assessing the Moisture Transfer of Your Outfit

In Ch. 3, we defined the water-vapor permeability index (i_m) as a dimensionless unit between 0 and 1, where 0 means no, and 1 means perfect exchange of moisture with the ambient air. The moisture exchange depends on the ambient water-vapor pressure and temperature (Fig. 4.1). Then values of i_m close to 1 would occur only with high wind speed, and no clothing. Recall, moisture transfer is a diffusion process that is limited by the characteristic vapor diffusion in still air. Consequently, in still air, most permeable clothing ensembles have i_m values less than 0.5.

Very tight weaves, water-repellent fabrics, and chemical protective impregnation can reduce the i_m-value notably. Nevertheless, even impermeable layers seldom result in i_m-values equal to 0. This fact is due to the internal evaporation-condensation process that occurs between the skin and inner surface of the impermeable layer. This process namely transfers some heat from the skin to the vapor barrier.

The evaporative heat transfer from the skin through the clothing and air layers to the ambient air depends on the ratio of the water-vapor permeability index to insulation ratio (i_m/clo). Therefore, we can estimate the maximum evaporative heat exchange, E_{max}, with the ambient air as [118]

$$E_{max} = 10\ i_m\ /\text{clo L}\ (T_{skin}\ e_{skin} - rh\ e_s).$$

Here L=2.2 is the *Lewis number*, which gives the ratio of thermal diffusivity (heat) to mass diffusivity (water-vapor). Again, e_{skin} is the water-vapor pressure of sweat at skin temperature T_{skin}, $rh=e/e_s$ is the fractional relative humidity given in values between 0 and 1, e and e_s are the actual water-vapor pressure and saturation water-vapor pressure at air temperature, T_a. Multiplication of the fractional relative humidity with 100 provides the relative humidity in percent.

The above equation tells us that the maximum evaporative transfer depends quasi-linearly, and inversely on insulation. In simple words, the evaporative transfer decreases with increasing insulation and vice versa. The **take home message** is that it is way easier to dress for cold-dry than cold-humid conditions or hot, humid conditions. As a result, to avoid feeling cold in winter, you will have to spend extra efforts on staying dry in humid climate or under humid weather conditions.

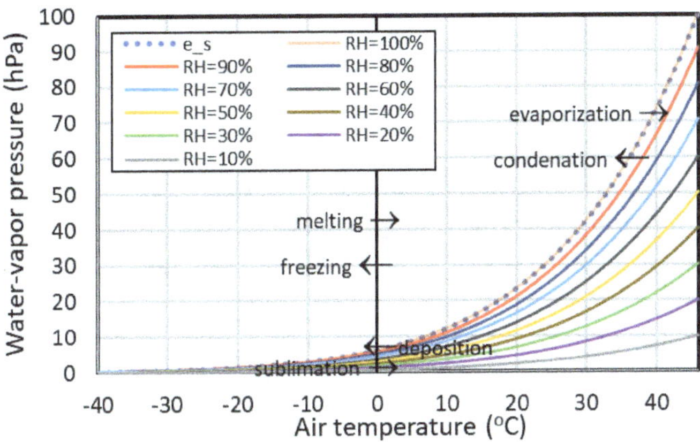

Fig. 4.1. Phase diagram of water. Water-vapor pressure, e, relative humidity, RH, and saturation water-vapor pressure, e_s, vs. air temperature. In the region below the saturation water-vapor pressure curve, water is in the gaseous phase as water vapor. The phase transition from the liquid to gaseous phase and vice versa are called evaporation and condensation. Below the freezing point, liquid water freezes, while above 32F (0°C), ice melts. Ice may go directly in the gaseous phase (*sublimation*), when the air is desaturated with respect to the saturation vapor-pressure over ice. At supersaturation with respect to ice, water vapor may directly go in the ice phase (*deposition*).

4.3 Impact of Your Activity on Clothing Insulation

Remember your mom's insisting on keeping good posture? Your posture and activity affect the clothing insulation. Sitting or lying down compresses the air layers in your outfit, and consequently changes its thermal insulation. Recall, your motion decreases the insulation of your outfit because it pumps air through openings and/or causes air to move within the layers leading to forced convection (Fig. 3.1c). The impact of motions on the insulation depends on both the kind of activity and clothing. Therefore, accurate estimates of clothing insulation during sportive and/or other activities, $I_{cl,active}$, hardly exist, but can be broadly estimated as

$$I_{cl,active}=I_{cl}\cdot(0.6+0.4/M) \qquad \text{for 1.2 met} < M < 2.0 \text{ met.}$$

Here M is the *metabolic rate* (in met). Note that 1 met is defined as 50 kcal of *metabolic heat production* per square meter of body-surface area per hour. Furthermore, I_{cl} is the insulation without activity as given in Tables 4.1 and 4.2. For metabolic rates less than or equal to 1.2 met, no adjustment should be made.

According to the *conservation of energy*, we can formulate an energy balance for the *system body-clothing-environment*. The energy fluxes that affect the body are the radiant fluxes from the Sun (Fig. 2.2), the reflected (diffuse) radiation, the IR-radiation emitted by the human body and obstacles in the environment (walls, ground, clouds, etc.), the IR-radiation (aka *long-wave radiation*) absorbed by the human body, heat conduction, convection, sensible heat fluxes, *latent heat fluxes* due to transpiration, evaporation, respiration (Fig. 4.2), and chemical energy. We can describe the interaction between our metabolic heat production, MH, and the energies that can change the body temperature, i.e., cause a change in *heat storage*, ΔS, as

$$MH=\Delta S\pm(R+Q_{conv}+Q_{cond})+L_vE.$$

Herein, Q_{conv}, Q_{cond}, R, and L_vE are the heat exchange by (free and forced) convection, conduction, (solar, diffuse, long-wave) radiation, and latent heat fluxes (evaporation, transpiration, respiration), respectively. Rearranging this equation, we can calculate the change in heat storage, ΔS. When this term is negative, our body produces insufficient heat to compensate for (i.e., balance) the heat loss to the environment, and our body-heat content decreases. If ΔS is positive, we will feel hot.

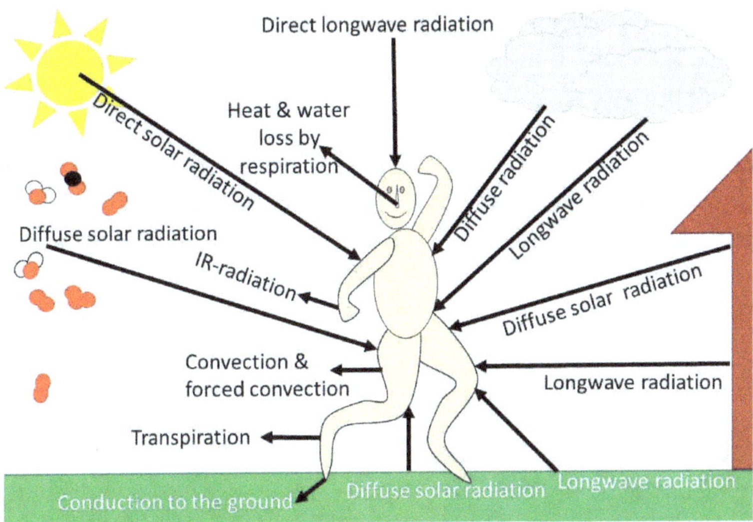

Fig. 4.2. Schematic view of the energy balance between the human body and its environment.

In cold weather, it is important to keep your heat loss as low as possible. The lower the loss, the longer you can stay outside without feeling uncomfortable and risking hypothermia. Because the mean specific heat capacity of human tissue is 0.83 kcal/(kg·K), a change in body temperature of 1.8F (1°C) for a 132 lb (~60 kg) person would mean a heat loss of 48.8 kcal. When MH is less than $R+Q_{conv}+Q_{cond}$, we have a heat debt.

At rest, our body-heat production is about 1 met. This means our body produces about 90 kcal/h (105 W) of heat assumed we have a mean body-surface area of about 19.4 ft^2 (1.8 m^2). This heat production is about twice and three times as high at normal and hard work[34], respectively.

In a comfortable environment at rest, we lose about 12% of our body-heat production by respiration, 12% by transpiration, and about 76% due to radiation, conduction and convection. The ambient air motion governs the partitioning between the heat loss by convection, conduction, and radiation. In calm air with minimal clothing, the loss

---

[34] Humans can sustain hard work at about 5 met (425 kcal/h or 500 W).

due to our body's IR-radiation is about 60% of the heat production at rest. Note that some of the elevated heat production during work is lost by forced convection due to the enhanced convective air motion caused by the body's motion. Even in cold weather, we may lose still about 42% of our body-heat production due to evaporation of skin moisture and sweat.

The optimum for clothing's overall thermal conductivity depends on the balance between body-heat radiation and convection. In simple words, the denser the fabric, the less body-heat loss occurs through the fabric, but the more body heat loss occurs through conduction.

Obviously, self-assessing clothing insulation and moisture management on the basis of garments has several shortcomings. The reasons are that the values of thermal conductivity, bulk density of fabrics, and moisture management of garments vary with the twist of the yarn, tightness of the knit or weave, fiber type and fiber blend ratio as well as the fabric moisture content (see Ch. 3). Unfortunately, garment labels only list the blend ratio – even tough without specification whether the blend was created at the fiber level or on the loom.

5 THE ROLE OF WEATHER CONDITIONS FOR THERMAL STRESS

5.1 Relative Humidity and Thermo-Physiological Comfort

The typical person from Anchorage, Alaska would not want to move to Fairbanks, Alaska because of the even lower air temperatures than at home. On the other hand, the typical Fairbanks person would not want to live in Anchorage because of the much higher humidity than in the dry Interior. The Fairbanks citizen would argue the humidity in the Cook Inlet is creepy, and makes the cold air feeling even colder than in Fairbanks. On the other hand, looking at hot weather, 100F (37.8°C) are more bearable in the dry Utah desert than in the humid air of Virginia at same wind and insolation conditions.

Let's take a step back to understand the reasons for the impact of relative humidity on thermal comfort under cold and hot conditions. Recall sweating serves to regulate the body temperature because evaporation results in cooling (Ch. 3). However, not all heat consumed by evaporation of sweat contributes to cooling of the body. Some of the cooling effect is lost to the environment. Therefore, the efficiency of sweating amounts only 59 to 71%.

Another aspect affecting the efficiency of sweating is that an increase in humidity is less than that of an *adiabatic*[35] change in humidity (Fig. 4.1). Consequently, evaporative cooling always reduces the *apparent* temperature[36]. However, an active person can only achieve comfort this way at apparent temperatures below 77.5F (25.3°C). Above this threshold, a change in water-vapor pressure by 10 hPa has the same effect as a temperature change of 3.4F (1.9°C; Fig. 4.1). This fact means that evaporation of sweat strongly decreases when the relative humidity in the environment increases. As a result, at high relative humidity under hot conditions, sweat can accumulate on the

---

[35] An adiabatic change means that there is no exchange of heat or mass with the environment. The internal energy and temperature change of an air parcel is solely due to work, i.e., compression or expansion of the air parcel. For an ideal gas and most atmospheric conditions, compression and expansion result in warming and cooling, respectively.

[36] You have made the experience of evaporative cooling when you leave the shower with wet skin on a non-muggy day.

skin (Fig. 3.10). In this case, the sweat forms an insulating water layer that hinders the heat transfer from the body to the ambient air. In addition, dripping of sweat further reduces the efficiency of sweating because the dripped sweat doesn't evaporate on the skin. Consequently, in hot, muggy weather with relative humidity close to or 100%, sweating fails to cool the body because barely any or no evaporation occurs. This means the humid air limits and slows down the amount of sweat evaporation, on hot, humid days. Therefore, you just feel sweaty, and perceive the air as hotter than it actually is. This higher apparent temperature is due to the combination of the actual temperature, and the high water-vapor pressure. People call this combination sultriness. In general, sultriness can increase by increases in ambient temperature, humidity, and solar radiation, as well as slowing down of wind speed or changes in atmospheric pressure.

Consequently, achieving thermo-physiological comfort on muggy days like the *Dog Days* is a challenge. A typical adult of 5 ft 6 (1.7 m) height and 147 lb (67 kg), for instance, has a surface area of 19.2 ft^2 (1.78 m^2), and a body density of 980 kg·m^{-3} (Tab. 3.1). An outfit of long pants, and a short sleeve shirt/blouse covers about 84% of the body surface. Under these considerations, the clothing thickness (includes underwear plus air layers) amounts about 0.2 inch (5 mm). According to medical literature, the core vapor-pressure of the human body is that of a saline solution with an effective relative humidity of about 90%. At 98.6F (37°C), which is the human blood temperature, the water-vapor pressure at saturation amounts 0.628 hPa (mb). Given the 90% relative humidity of the human body, our body-vapor pressure is about 0.56 hPa (mb). It is obvious that the difference between the two values is very small. Consequently, evaporation of sweat is low, and hence, fails to provide the much-wanted cooling.

On the contrary, in cold climates, evaporative cooling is unwanted. Recall, at high relative humidity, cool/cold air feels colder than it actually is. Under these conditions, sweat may re-condense in the outer layers of the clothes (cf. Fig. 4.1). As a result, evaporative cooling occurs in the clothes. Obviously, this physical process is a problem when working out outside in winter.

As discussed in Ch. 3, the clothing and thin air layers in between prevent your body from losing heat by convection. On cold, humid days, this air is humid too. Your body needs more heat energy to warm moist than dry air due to the higher specific heat capacity of water than air (Tab. 3.1). Consequently, when your skin is in contact with a layer of damp air, your body loses more heat energy to that air layer. Therefore, you perceive the air to be creepy and cooler than it actually is. Furthermore, sweat (water) conducts heat better than air does (cf. Tab. 3.1). Consequently, when the dampness leads to traces of liquid water on your skin or in your clothing, your body experiences an increased heat loss by conduction.

At temperatures below 32F (0°C), water freezes, and may later sublimate (Fig. 4.1). During freezing, latent heat is released, which has a warming effect. However, once sublimation kicks in, heat is consumed, which leads to a cooling that corresponds to the sum of the cooling by melting and evaporation. Therefore, in cold climate, it is inevitable to ensure your clothing stays dry.

When it is cold, the body does not need sweating as a cooling mechanism. Despite you perspire less in the cold, you may still perspire, especially, when you perform an energy-consuming outdoor activity. Like humidity in hot climate, the humidity limits the evaporation of sweat. However, in cold weather, the sweat cools down, which also contributes to perceiving the air as colder than it actually is. Furthermore, on cold damp days, the clothing absorbs moisture and may become wet, which reduces the insulating factor of the clothing[37].

At temperatures below 32F (0°C), the water/sweat in wet or damp clothing freezes and makes your clothing stiff like lumber. At temperature below -40F (-40°C) even hot water instantly freezes when exposed in low amounts to the air[38].

5.2 Rain and Rainy Weather

On a cold rainy day, when you do not wear raingear, rain soaks your clothing. As discussed already, the wet clothing make you feel colder due to evaporative cooling and reduced insulation. Furthermore, because raindrops form at higher, colder levels in the atmosphere, they are typically cooler than the air temperature close to

---

[37] You probably made yourself the experience that you lose heat quickly, and get (unwanted) cooling when your swimsuit is wet.

[38] On 40 below days, a favorite demonstration given to newcomers to the Arctic is to throw a steaming hot coffee into the air. Guess what? The coffee droplets immediately freeze. Another nice demonstration is to make soap bubbles. They immediately freeze, and shatter like porcelain once they hit the ground.

the ground. They namely fall too fast (up to 22.4 mph, 10 m/s) to adjust fully to the new ambient air temperatures. Consequently, the cool rain also contributes to thermal discomfort.

5.3 Wind and Thermal Discomfort

When it is windy, air feels colder on the skin than under calm wind conditions. Wind, namely, increases ventilation, leading to forced convection and enhanced heat loss (Fig. 3.1). In general, a surface loses heat at a rate proportional to the wind speed above this surface. The stronger the wind, the more quickly the surface cools[39]. This phenomenon is known as *wind chill*. The wind chill explains why in Fairbanks, on a calm day, -40F (-40°C) air temperature feels less bad than an air temperature of -10F (-23.3°C) with 60 mph (26.82 m/s) wind on an Aleutian Island.

Wind chill is defined as the air temperature felt on exposed skin due to wind. Consequently, the *wind-chill temperature* remains always lower or equal to the actual air temperature. Note that wind chill is not defined for temperatures above 50F (10°C) and wind speeds less than 3 mph (1.34 m/s).

Wind chill reduces the temperature of any warmer objects than the ambient temperature more quickly than these objects would adjust to the ambient temperature otherwise. A healthy person's body responds to wind chill by burning more calories. This metabolism behavior serves to maintain the skin temperature in an acceptable range. As a result, a healthy person needs more chemical energy in form of food or the body burns body fat.

This fact should not give you any wrong ideas about losing weight by not dressing appropriately. As a rule of thumb, a scientist needs about 4000 kcal more per day when pursuing fieldwork on the sea ice in Utqiagvik (formerly Barrow), Alaska, in winter than in summer. While just eating more to provide the energy needed may be tempting and delicious, doing so will not protect you from frost-nip, frostbite, hypothermia, and other health-adverse cold-related effects or even death. Therefore, you should strive to maintain a given skin-surface temperature in a cold environment, especially when there is increased heat loss due to wind chill. In simple words, you should protect your

---

[39] Recall your mother's blowing your hot food when feeding you with a spoon when you were a small kid.

body from the elements with clothing as best as you can. This means that to dress appropriately in windy cold climate, you have to know that wind chill increases with decreasing air temperature.

Because no universally agreed standard exists, you find various equations for determining wind chill. All these equations strive to qualitatively describe the temperature a person will perceive due to the heat loss by the wind. Fortunately, the national weather services of the different countries use one standard for their country. As a result, you can develop an experience with the *wind-chill system* used in the country you live. However, once you visit a country using a different standard, your experience may be useless. Therefore, to protect yourself, it may be wise to get the equation used in your country. Then you can calculate the wind-chill temperature that matches your experience. I bet there is even an app out there to do so.

When performing field work in Antarctica, Paul Allman Siple and Charles Passel developed the first wind-chill equations and -tables. The US National Weather Service started using them in the 1970s. Allman and Passel's *wind-chill index* is based on the cooling rate of a small plastic water bottle as the water froze while hanging in the wind on the roof of the expedition hut at the same level as the anemometer.

Initially, the units of the wind-chill index sounded very strange to the public. The public was uncomfortable with the wind-chill index and preferred a *wind-chill equivalent temperature* (WCET) like it was introduced in the 1960s. Back then, the WCET was defined as the temperature at which the wind-chill would be equal to the temperature in the absence of wind. This definition of the WCET obviously exaggerated the severity of the weather. Furthermore, even when winds are calm, air still moves. To obtain more realistic sounding WCETs Charles Eagan redefined the term "absence of wind" to be a wind speed of 4 mph (1.8 m/s).

Medical experts and meteorologists developed a new wind chill index by means of a model that used standard engineering correlations of the wind speed and heat-transfer rate. The model iteratively determines the skin temperature under various temperatures and wind speeds. It calculates the heat transfer for a bare face facing the wind, while walking into the wind at 3.1 mph (1.4 m/s). Because according to the World Meteorological Organization (WMO) standards, wind is typically measured at 10 m height, the model corrects the officially measured wind speed to the wind speed at face height.

In 2001, the United States and Canada introduced this new wind-chill index. Environment Canada uses the following equation to calculate their wind-chill index in °C

$$T_{\text{wind-chill,C}} = 13.12 + 0.6215 \cdot T_{\text{air}} + (0.3965 \cdot T_{\text{air}} - 11.37) \cdot v_{\text{10m}}^{0.16}.$$

Here T_{air} is the actual air temperature in °C measured at 2 m (2.19 yd) height, and v_{10m} is the wind speed in km/h measured at 10 m (10.94 yd) height. The wind-chill equation applied by the US National Weather Service (NWS) reads

$$T_{\text{wind-chill,F}} = 35.74 + 0.6215 \cdot T_{\text{air}} + (0.4275 \cdot T_{\text{air}} - 35.75) \cdot v_{\text{10m}}^{0.16}$$

Where the temperature and wind-speed input data are in Fahrenheit and mph, respectively, and $T_{\text{wind-chill,F}}$ is in degree F. Both equations are valid only for wind speeds exceeding 4.8 km/h (3 mph) and air temperatures below or equal to 50F (10°C).

This 2001-WCET is a steady-state value except for the estimated time to frostbite. However, wind chill has notable, time-dependent, i.e., non-steady aspects to it. The cooling is most rapid at the beginning of any exposure, when the skin is still warm. At that time, the temperature difference between the skin and air is namely the largest. In the development of the wind-chill equation, it was assumed that a person wears appropriate dry clothing. This means the equation ignores 1) the actual type of clothing worn, and 2) effects related to wet clothing and/or skin.

Scientists have discussed the methods for estimating wind chill controversially. They mainly disagree about whether or not to consider a naked or an appropriately dressed body, and whether or not to consider the cooling of the entire body, the face or exposed skin. Furthermore, they disagree about whether or not, and if, how to consider the large variability in the internal thermal resistance among persons.

Independent of the scientific dispute, all that matters for you is to protect your body from wind chill by wearing wind-proof clothing when it is windy. When your body is losing too much heat you may be in danger. Your body, namely, reacts for the sake of survival. Therefore, it may shut down the blood flow to body parts that are "expendable" and/or loose the most heat. This natural process can result in frost-nip, frostbite, and finally the loss of that body part. Obviously, fingers and toes are most endangered.

Temperature (°C)

Wind speed (km/h)	5	0	-5	-10	-15	-20	-25	-30	-35	-40	-45	-50	-55
10	3	-3	-9	-15	-21	-27	-33	-39	-45	-51	-57	-63	-69
20	1	-5	-12	-18	-24	-30	-37	-43	-49	-56	-62	-68	-75
30	0	-6	-13	-20	-26	-33	-39	-46	-52	-59	-65	-72	-78
40	-1	-7	-14	-21	-27	-34	-41	-48	-54	-61	-68	-74	-81
50	-1	-8	-15	-22	-29	-35	-42	-49	-56	-63	-69	-76	-83
60	-2	-9	-16	-23	-30	-36	-43	-50	-57	-64	-71	-78	-85
70	-2	-9	-16	-23	-30	-37	-44	-51	-58	-65	-72	-80	-87
80	-3	-10	-17	-24	-31	-38	-45	-52	-60	-67	-74	-81	-88
90	-3	-10	-17	-25	-32	-39	-46	-53	-61	-68	-75	-82	-89
100	-3	-11	-18	-25	-32	-40	-47	-54	-61	-69	-76	-83	-90

(a)

Temperature (F)

Wind speed (mph)	40	35	30	25	20	15	10	5	0	-5	-10	-15	-20	-25	-30	-35	-40	-45
5	36	31	25	19	13	7	1	-5	-11	-16	-22	-28	-34	-40	-46	-52	-57	-63
10	34	27	21	15	9	3	-4	-10	-16	-22	-28	-35	-41	-47	-53	-59	-66	-72
15	32	25	19	13	6	0	-7	-13	-19	-26	-32	-39	-45	-51	-58	-64	-71	-77
20	30	24	17	11	4	-2	-9	-15	-22	-29	-35	-42	-48	-55	-61	-68	-74	-81
25	29	23	16	9	3	-4	-11	-17	-24	-31	-37	-44	-51	-58	-64	-71	-78	-84
30	28	22	15	8	1	-5	-12	-19	-26	-33	-39	-46	-53	-60	-67	-73	-80	-87
35	28	21	14	7	0	-7	-14	-21	-27	-34	-41	-48	-55	-62	-69	-76	-82	-89
40	27	20	13	6	-1	-8	-15	-22	-29	-36	-43	-50	-57	-64	-71	-78	-84	-91
45	26	19	12	5	-2	-9	-16	-23	-30	-37	-44	-51	-58	-65	-72	-79	-86	-93
50	26	19	12	4	-3	-10	-17	-24	-31	-38	-45	-52	-60	-67	-74	-81	-88	-95
55	25	18	11	4	-3	-11	-18	-25	-32	-39	-46	-54	-61	-68	-75	-82	-89	-97
60	25	17	10	3	-4	-11	-19	-26	-33	-40	-48	-55	-62	-69	-76	-84	-91	-98

Time to frostbite: 30 min   10 min   5 min

(b)

Fig. 5.1. Wind-chill table in (a) metric units, and (b) British units. The coloring gives the time it takes to get frost-bite under a given temperature and wind combination. The temperature values in the charts give how you would feel the temperature. To determine the wind-chill temperature, find where the actual temperature- and wind-speed lines cross.

Your food affects how well your body retains its heat. In avoiding hypothermia, being well hydrated is important according to medical studies. Therefore, the National Institute of Health recommends to drink plenty of water prior, during and after your winter-outdoor activities; take warm drinks if possible, and consume enough calories to maintain the needed energy levels during outdoor activities. In cold weather, you might burn more than 6000 kcal a day when performing high energy activity.

5.4 The Extremes - Heatwaves and Cold-Snaps

Blocking high pressure systems can cause *heatwaves*. Heatwaves, especially, when combined with *urban heat-island* effects and/or high humidity, can lead to heat stress, thermo-physiological discomfort, health issues and even death. Heatwaves are an extended period (>4 days) of hot weather relative to the weather conditions with mean temperatures expected to persist over four days once in 10 years based on the *local historical records*. This means different locations have different temperature thresholds for heatwaves because of their different historical records.

Cold-snaps aka *cold-waves* or *cold-spells* are an unusually large, rapid drop in temperature over 24 h. This means they are independent of historic observation and likelihood of occurrence. Cold-snaps occur when Arctic/Antarctic or Polar airmasses suddenly move towards lower latitudes. Cold-snaps lead to cold stress, thermal discomfort, health issues, and maybe death. In contrast to heat-waves, cold-snap related health issues and the related risk of death may persist for weeks after the event.

5.5 Radiative Energy Exchange between the Human Body and Ambient Air

Recall Planck's law that every object including air radiates at its own temperature. Cold bodies radiate with lower energy than hot bodies. In accord with the 2nd law of thermodynamics (Ch. 1), net energy flows from the higher to the lower energy; or in simple words, from the warmer to the cooler body. Consequently, a dressed person and their exposed skin loose energy by IR-radiation when they are warmer than the ambient air. In this case, they act as a heat source. On the contrary, when the clothes and body are colder than the ambient air, they gain energy, and their temperature rises.

Now let's translate this concept into the fashion world. There will be no loss by radiation when your body, clothes and the air have the same temperature. However, even when you dressed appropriately, your body will lose heat when the environment is colder than your body temperature. The larger the difference between the surface temperature of your outerwear, and the environment is, the higher is the loss of radiation energy. In general, the heat flux by radiation at the outer surface of the clothes reads

$$Q_{rad} = \varepsilon\sigma(T_{clothes}^{\;4} - T_{air}^{\;4})$$

Where $T_{clothes}$ and T_{air} are the mean temperature at the outer surface of the clothes and the mean air temperature at the clothes-atmosphere interface in Kelvin (see Ch. 1). Furthermore, ε is the emissivity of the garment which depends on the fabric conditions (e.g., color, structure), and $\sigma = 5.67\cdot10^{-8}$ W m^{-2} K^{-4} is the *Stefan-Boltzmann constant*.

Often, an empirical formula serves to calculate the heat loss due to radiation from the outer surface of a textile assembly

$$Q_{rad} \cong 3.97\cdot10^{-8}\cdot T_{clothes}^{\;4} - T_{mrt}^{\;4}\cdot A_{Du}\cdot f_{clothes}$$

Where T_{mrt} is the *mean radiant temperature* in Kelvin; A_{Du} is the *Du Bois surface area*[40], and $f_{clothes}$ is the ratio of the effective radiating surface to the Du Bois surface area.

Now, let's look at radiation in terms of heat flow. Electromagnetic radiation transports the heat between the environment and clothing surface as well as between clothing layers. Furthermore, radiant heat transport can occur between the fibers within a textile through the entrapped air. The more fibers, the less radiant heat transfer occurs. In all cases, the temperature difference between the heat emitter and heat absorber governs the magnitude of the *radiation-heat flow*.

Looking at the role of clothes for thermal comfort this energetic concept means the following: Because a body radiates energy corresponding to its emissivity and temperature, our skin emits heat (thermal radiation) of wavelengths between 700 and 1400 nm with a peak at 950 nm. Recall, these wavelengths are in the IR-region (Fig.

---

[40] In medicine and physiology, various equations exist to access the body surface area. Therefore, the term Du Bois surface area indicates that the *Du Bois formula*, $A_{Du}=0.20247\cdot$Height$^{0.725}\cdot$Weight$^{0.425}$, served to calculate the body-surface area. Here the units of height, weight, and the Du Bois surface area are m, kg, and m^2, respectively.

2.2). This IR-radiation contributes more than 50% of our total body-heat loss[41]. Consequently, the infrared optical properties of fabrics are key in achieving a required cooling/heating demand in advanced personal thermal wear (see Ch. 8). This means fabrics require not only the right UV- (Ch. 2), but also the right IR-optical properties.

According to *Kirchhoff's law*, the energy balance of any thermal emitter is

$$\varepsilon_\lambda + \tau_\lambda + \varrho_\lambda = 1$$

Where ε, τ, and ϱ are its emissivity, transmittance, and spectral reflectivity, respectively, at the wavelength, λ.

Let me explain Kirchoff's law in terms of clothing. The infrared emissivity of our body is 0.98. To obtain maximal cooling under hot conditions, clothing should permit high dissipation of your body radiation. This requires a fabric emissivity of near 1, or high transparence within the mid-infrared wavelengths. On the contrary, in cold weather, the fabric's reflectivity should be close to 1 in the mid-infrared range to achieve heating. However, for both cooling or warming purposes, fabrics should be opaque in the visible light spectrum (Fig. 2.2).

Radiation effects also contribute to perceiving cold weather as colder under humid than dry conditions. On clear sky, sunny days, radiative heating from the Sun warms your body, while on cloudy or even overcast days the direct solar radiation reaching you is small. Consequently, cloudy days feel cooler than sunny days when all factors (wind, temperature, humidity) remain the same.

5.6 Thermoregulation of Our Body

As discussed in Ch. 4, our body exchanges heat with its environment via conduction, convection, radiation, and transpiration (Fig. 4.2). At all times, we lose latent heat (water vapor) from the lungs due to breathing. To compensate for heat losses, our body continuously produces heat via metabolic reactions. At rest, our internal organs produce the most heat. When active, our skeletal striated muscles can contribute up to 90% of the overall heat

---

[41] In the so-called *atmospheric window* (800 nm – 1400 nm; Fig. 2.2), this thermal radiation can dissipate into space.

production. Activity like exercising for 10-20 minutes, for instance, can rise blood temperature to 100.4F (38°C), and increase sweating [121]. When walking outdoors at 1.4 m/s, the skin surface puts out about 180 J s^{-1}m^{-2} of heat by transpiration. In addition, walking accelerates the heat exchange between your body, the air in your clothing layers and the ambient environment. Being in the shadow or in the Sun can influence the metabolism, and makes a huge difference in thermal comfort.

5.6.1 Thermal Heat and Cold Stress

As discussed in Ch. 4, the energies can change the body temperature; the human metabolism and behavior can adapt to these conditions within limits for thermal comfort. Once the environmental conditions are outside our thermal comfort zone, we experience thermal stress.

High levels of heat stress, or cold stress can impact a person's health, and reduce the efficiency of performed activities. Under long-term exposure both heat and cold stress can cause death. Heat stress varies among regions depending on whether extreme moisture or extreme temperature dominate. During heatwaves, mortality raises for several days. The theoretical limit, at which humans would die due to heat stress after 6 h of exposure, is a relative humidity of 100%, and air temperature of 95F (35°C) at the same time. Fortunately, such conditions rarely occur for such long.

The metabolism of a person exposed to extreme cold ambient air reduces the blood flow in their skin to avoid body heat loss. That's why the skin gets cold. Death due to hypothermia can already occur at temperatures close to the freezing point. Cold stress is a high-risk factor among patients, particularly women, with cardiovascular diseases [122]. As aforementioned, cold-spells increase mortality for several weeks [123] because more people catch a cold, the flu or pneumonia than under normal winter conditions.

5.7 Thermal Stress Indexes

Typically, so-called *thermal stress indexes* express the impacts of thermal stress on human health and performance. Scientists developed thermal stress indexes that differ, among other things, with respect to their concept (empirical, direct, indirect, analytical), application goals, the considered meteorological quantities (e.g., airflow velocity, temperature, humidity, solar radiation), and whether or not clothing is assumed, and if, the kind of clothing.

Empirical thermal stress indexes base on human comfort or responses to different environmental factors. On the contrary, direct indexes use measurements. The so-called *apparent temperature*, for instance,

$$AT=T_a+0.33 \cdot v_a-0.7 \cdot v_a^{-4}$$

Uses the observed ambient air temperature, T_a, and wind speed, v_a. It combines the heat index and wind chill effect. The most commonly used thermal stress index is the *Wet-Bulb Globe Temperature*

$$WBGT=0.567 \cdot T_a+0.393 \cdot v_a+3.94.$$

So-called *analytical indexes* consider the principles of thermal heat and moisture exchange between the human body and its environment. The *Standard Effective Temperature* (SET), for instance, considers physiological parameters (skin temperature, wetness).

Depending on the geographical location, different aspects are in the focus of describing the stress conditions. The *heat index*, and *humidex*, for instance, are only valid for ambient air temperatures above 68F (20°C), while the wind-chill index, and wind-chill equivalent temperature index are valid only at and below the freezing point (32F, 0°C). The *Mediterranean Outdoor Comfort Index* (MOCI) was optimized for Mediterranean regions [124].

Other thermal stress indexes, for instance, the *Effective Temperature* and the *Physiological Equivalent Temperature* consider the energy balance, but these indexes apply only for indoor activity. Some indexes originally developed for indoor conditions were extended to outdoor conditions by including the effects of solar radiation [125], [126].

Research on the impacts of weather on health showed that the mean radiant temperature, T_{mrt} better describes thermal stress than the ambient air temperature [127], [128]. It namely considers, among other things, the body posture (sitting, standing, etc.), shortwave and long-wave radiation fluxes from all sides, the body's absorption of shortwave and long-wave radiation, and wind speed. The mean radiant temperature is defined as the "uniform temperature of an imaginary enclosure in which the radiant heat transfer from the human body equals the radiant heat transfer in the actual non-uniform enclosure"

[118]. In simple words, T_{rmt} represents the combined radiant and convective heat gains/losses that affect the human energy balance outdoors (Fig. 4.2). Consideration of these factors is important because T_{mrt} of shaded and sunlit areas can differ up to 66.6F (37°C) according to studies performed in Freiburg, Germany [129]. At the same time, the air temperature, T_a, between shaded and sunlit areas differed only 1.8-3.6F (1-2°C).

An example of a thermal stress index that considers T_{mrt}, is the *Global Outdoor Comfort Index* (GOCI) [130]. This empirical index depends on T_a, relative humidity, v_a, geographic latitude, the mean annual temperature, and the mean temperatures of the hottest and coldest months.

Most thermal stress indexes assume a standard outfit consisting of a light shirt, and slacks. Even a layman can tell that such an outfit is unsuitable for many weather conditions worldwide.

5.7.1 Universal Thermal Comfort Index

Because of the short-comings related to no consideration of appropriate clothes for the actual environmental conditions, researchers organized an international workshop. Within this framework, the participants decided to develop a thermal stress index that encompasses adequate clothing, human behavioral responses, human metabolism, and the energy balance between the dressed person's body and its environment. This *Universal Thermal Comfort Index* (UTCI) considers a clothing model including behavioral responses [131], a human-metabolism model [132], the energy balance for the system *body-clothing-environment*, as well as empirical relations between UTCI (expressed as a temperature), and the thermal comfort experienced by test persons (Fig. 5.2). As a result, the UTCI combines thermo-physiology, and heat-exchange theory while following the concept of an equivalent temperature.

At any combination of air temperature, wind speed, radiation, and humidity, the UTCI is defined as the air temperature, T_a, in the reference condition, which would elicit the same dynamic physiological response as the actual conditions. These reference conditions assume a metabolism of 135 $W \cdot m^{-2}$, light activity (walking 2.5 mph, 4 km/h), a mean radiant temperature being equal to the air temperature, a relative humidity of 50% for ambient air temperature below or equal to 84.2F (29°C), a water-vapor pressure of 20 hPa for ambient air temperatures above 84.2F (29°C), and a wind speed, v_a of 1.1 mph (0.5 m/s) at 10 m height[42], which corresponds to about 0.7 mph (0.3 m/s) at 1.1 m above ground. All other weather conditions are compared to this reference.

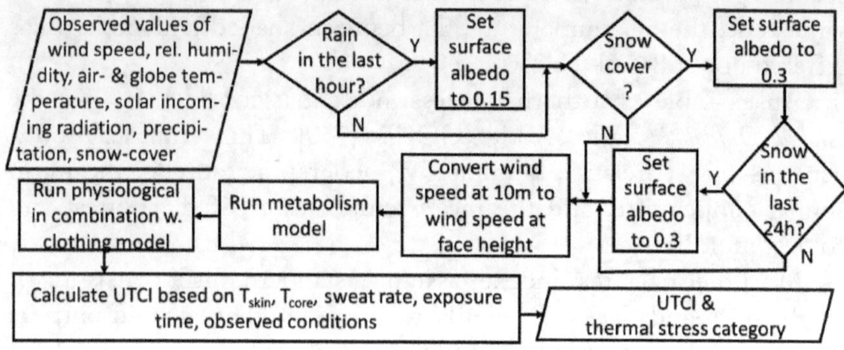

Fig. 5.2. Schematic view of the modified UTCI model [135].

The calculation of the UTCI bases on the advanced *multi-node human thermoregulation model* [132]. This computer model simulates, among other things, the environmental heat exchanges, the heat and mass transfer with the body (Fig. 4.2), thermoregulatory reactions of the central nervous system and perceptual responses. The well evaluated advanced multi-node human thermoregulation model is valid for ambient air temperature from −58F to 112F (−50°C to 50°C), differences between the mean radiant temperature and ambient air temperature (T_{mrt}–T_a) of −22F to 158F (−30°C to 70°C), wind speeds between 1.1 mph and 67.8 mph (0.5 m/s and 30.3 m/s), 5 to 100% relative humidity, and water-vapor pressure less than 50 hPa.

Recall, *thermoregulation* refers to the human's ability to keep the body temperature within certain limits despite of a very different ambient temperature. When the body-core temperature and/or skin temperature change, so-called *effector responses* occur; effector actions depend on weather conditions, clothing, and activity. Remember, in moderate and warm environments, a person's physiology can maintain body temperature within acceptable limits via thermoregulation; the body responses with sweating and vasodilation of skin vessels to cool the skin by evaporation of sweat (Fig. 3.10). In cold environments,

---

[42] In meteorology, wind speed and air temperature are measured in 10 m (10.94 yd) and 2 m (2.19 yd) height, respectively.

increased convection is the main path to body-heat loss. The thermoregulatory system reacts with a reduction of the peripheral blood flow. As a result, skin temperature decreases thereby increasing the thermal insulation of the skin tissue. Shivering may occur to produce heat [134].

Adequate clothing is a *behavioral response* to the thermal conditions. The clothing model [131] simulates people's adjusting what they wear under the given outside conditions. Consequently, like in real life, in the clothing model, the thermal insulation due to clothes alters with ambient air temperature. These thermal insulation values stem from data of the actual clothing behavior of people for the respective conditions; below -4F ($-20°$C), the clothing model assumes special clothing [131].

The clothing model considers a so-called *partial clothing approach*. In simple words, it accounts for the face's exposure to the ambient air, and the different covering of head and hands compared to the torso. In addition, it considers that both a person's movement, and wind reduce the thermal and evaporative clothing resistances. At ambient temperature, the clothing stays the same, while radiation, relative humidity, and wind modulate the physiological response.

The advanced multi-node human thermoregulation model [132] runs in combination with the adaptive clothing model [131]. It predicts the dynamic thermal sensation responses to the steady-state and transient conditions based on the physiological states (skin temperature, core temperature, sweat rate, skin wetness, etc.) provided by the metabolism model (Fig. 5.2).

The calculation of the energy balance between the adequately dressed person and its environment requires consideration of the environmental surface albedo and emissivity. In its original version, the UTCI model considered a grass surface. The modified version [135] considers the ground surface as grass until onset of a snow cover; the albedo for reflection of solar radiation is set to 0.2 and 0.15 for dry and wet grass, respectively, where wet grass is assumed on days with at least one rain event. In case of snow, a surface albedo of fresh (0.55), and aging snow (0.3) is used for days with and without snowfall, respectively. The calculation of the energy balance considers snow cover based on snow depth and/or snowfall observations. When no such data exist, the modified UTCI model assumes a closed snow cover at geographic latitudes north of 60°N between November and end of March.

Recall, the result of the UTCI-model is an equivalent temperature, the UTCI (Tab. 5.1). The difference between the UTCI-value and air temperature, T_a, depends on the actual air and mean radiant temperature, T_{mrt}, wind speed, v_a, and moisture (water-vapor pressure, e, or relative humidity, RH in %). The model converts the wind speed observed at 10 m to body level assuming a logarithmic vertical wind profile.

The mean radiant temperature in °C is calculated as [136]

$$T_{mrt}=[(T_g+273.15)^4+1.06\,10^8\cdot v_a^{0.58}\cdot(T_g\text{-}T_a)/(\varepsilon\cdot D^{0.42})]^{0.25}-273.15$$

Where T_g, v_a, T_a, ε, and D are the observed globe temperature, wind speed converted to body level, air temperature, emissivity (0.95), and diameter of the globe thermometer.

5.7.1.1 Relation of Thermal Comfort Indexes to People's Sensation

Relating thermal comfort index values to the people's sensation requires large surveys and experimental studies under well-defined, repeatable conditions. Unfortunately, weather conditions are never exactly the same. However, repeatable conditions are important to compare different clothing types, materials, layering, etc., and also to expose various test persons to exactly the same conditions under which the scientists collected their data.

Recall, the goal was to relate the environmental conditions to the comfort level of appropriately dressed persons. Therefore, researchers exposed their appropriately dressed test persons for a given amount of time (typically 1 h) to various sets of temperature, moisture, and wind speed combinations in controlled-climate chambers. Thereafter, they asked them about their thermal comfort to establish an empirical relationship between the combined conditions of temperature, wind, moisture and thermal stress (Tab. 5.1).

As previously said, people feel most comfortable in the temperature range of 48.2 to 75.2F (9 to 24°C). Values between 64.4 and 78.8F (18 and 26°C) fall into the so-called *thermal comfort zone*. At temperatures below 48.2F (9°C) a person can stay outside for an extended amount of time while still feeling comfortable when wearing outerwear with the right insulation. However, by no means, clothing could protect a person from hypothermia when they stayed outside for an unlimited amount of time at air temperatures below the freezing point.

Table 5.1. Thermal stress categories of the Universal Thermal Comfort Index (UTCI).

UTCI range (°C)	UTCI range (F)	Thermal stress category
UTCI>46	UTCI>114.8	Extreme heat stress
38<UTCI≤46	100.4<UTCI≤114.8	Very strong h. stress
32<UTCI≤38	89.6<UTCI≤100.4	Strong heat stress
26<UTCI≤32	78.8<UTCI≤89.6	Moderate heat stress
9<UTCI≤26	48.2<UTCI≤78.8	No thermal stress
0<UTCI≤9	32<UTCI≤48.2	Slight cold stress
−13<UTCI≤0	8.6<UTCI≤32	Moderate cold stress
−27<UTCI≤−13	-16.6<UTCI≤8.6	Strong cold stress
−40≤UTCI≤−27	-40≤UTCI≤-16.6	Very strong c. stress
−40>UTCI	-40>UTCI	Extreme cold stress

Interestingly, depending on the weather conditions, the air can feel warmer or colder than it actually is. When using the UTCI-modeling system to examine the thermal comfort under the hourly observed weather conditions in Alaska's different climate zones [135], I found that strong thermal heat stress can even occur in Interior Alaska (Fig. 5.3c).

Obviously, in Fairbanks Alaska, the air felt warmer than the actual daily mean air temperature on most days of the 2015 warm season (Fig. 5.3a). The same is true when looking at the monthly means of air temperature and the UTCI (Fig. 5.3b). Except April, 2015 cold season monthly means of air temperature were marginally lower than those of the UTCI. April is the melting and breakup season. The associated evaporation of meltwater and sublimation of snow increase the relative humidity of the air, while decreasing the near-surface air temperature. Consequently, the air temperature is perceived as colder than it actually is, which is indicated by the lower UTCI than T. Note that data of T and UTCI of other years show similar results.

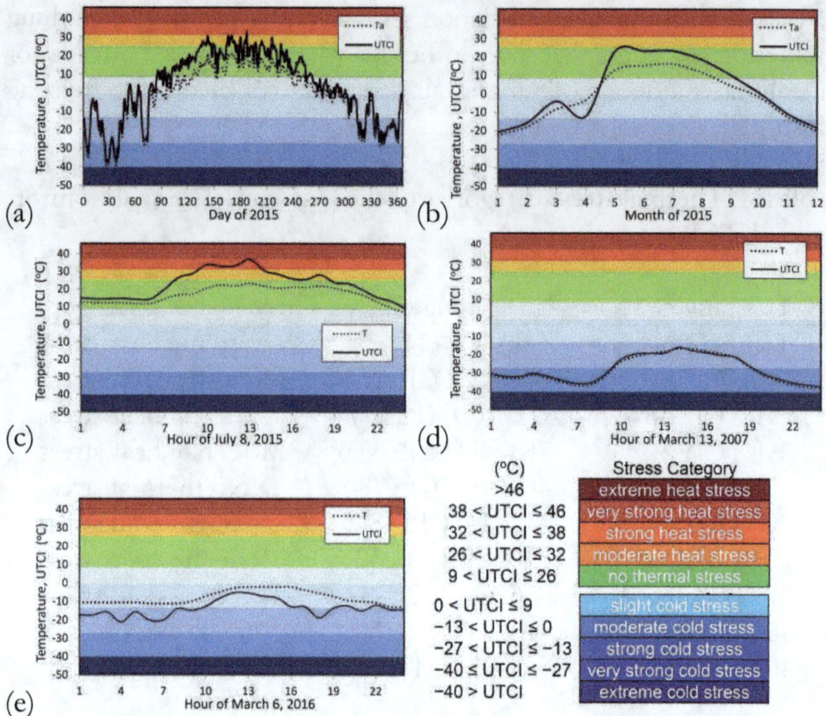

Fig. 5.3. Observed air temperature, T, modeled UTCI, and thermal stress category (color) for Fairbanks summarized as (a) monthly means in 2015, (b) daily means in March 2015, and hourly means in the diurnal course on (c) July 8, 2015, (d) March 13, 2007, (e) March 6, 2016. From [135].

It is important to realize that even within the diurnal course, there can be times where the UTCI exceeds the T, and vice versa (e.g., Fig. 5.3d). A decrease of wind speed and/or relative humidity and/or increase of solar radiation reaching the near-surface air, for instance, can lead to perceiving the air as warmer than before. On the contrary, increases in wind speed and/or relative humidity and/or decreases in solar radiation reaching the near-surface air aggravate the cold stress. Such changes in wind speed, insolation and relative humidity are typically associated with frontal passages or local weather phenomena (Ch. 7).

Obviously, dressing appropriately on days with changing weather conditions is difficult. Layering is your best approach to achieve thermal comfort due to the option to adjust the outfit. Knotting a sweater or cardigan around the neck or waist is a stylish way to carry the coverup needed when it feels colder later.

In a nutshell, the UTCI considers the impacts of wind, relative humidity, insolation, and temperature as well as the human metabolism under the assumption that the person is appropriately dressed for the actual weather conditions. The UTCI is expressed as a temperature that measures how we feel the temperature under the respective environmental conditions.

6 COLOR, FASHION, AND WEAR COMFORT

6.1 Color

Colors can vary in hue (cool or warm), chroma (light or dark), and value (muted/soft to bright/clear). It is believed that the human eye is able to distinguish about 10 million color variations. This means you have to pay attention to color when building your wardrobe to not look like you dressed in the dark. If you are a designer, paying attention to colors is an urgent need to create a sellable collection.

Dividing the spectrum into colors linguistically, and in fashion differs among cultures, and in history. Nevertheless, people seem to perceive colors in the same way. Typically, this perception is red, orange, yellow, green, blue, and violet. The intensity may alter the perception. As a result, a yellow-green or yellow-orange with low intensity may look olive and brown, respectively. Most languages distinguish six basic colors black, white, red, green, blue, and yellow. All languages use two basic color names to distinguish dark/cool and bright/warm colors. Together, these distinctions result in 12 colors: Black, gray, white, pink, red, orange, yellow, green, blue, purple, brown, and azure. In addition, many color names are derived from objects in this color, such as orange, salmon, lemon, oxblood, burgundy, chocolate, lilac, sage, olive. This principle also applies to further-distinguish a color, e.g., eggnog yellow, dandelion yellow, lemon yellow, or grass green, lime green, forest green, mint green.

6.2 Is White or Black Better for Summer?

Many people believe that wearing white in summer would keep them cooler than wearing any other color because white reflects the most sunlight. However, we have to consider the entire system *Sun-clothing-body* because our body emits energy in form of body heat in the IR-range. This means due to the high albedo of white, the inside of white garments also reflects the body heat back to your body. Let's take a close look at the two extremes – black and white.

Around the Mediterranean Sea and in the Sahara Desert, many people favor wearing black clothing. A study investigating the reason for this behavior found that the Tuareg[43] study participants felt more comfortable in dark than in white apparels [137]. During the study, the persons' body produced an average total heat energy of 512.9 Wm^{-2}, and the measured solar radiant energy was 857.3 Wm^{-2}. As we learned in Ch. 2, the reflection from a surface in the visible range can differ strongly from that in the infrared range. The reflection on the outside of the clothes occurs in the visible range of the solar spectrum (Fig. 2.2), while on the inside, the reflection of the body heat occurs in the infrared range. In contrast to white, black absorbs body heat. In the study, the dark and white apparel reflected about 84% and 57% of the solar radiant energy, while allowing 16% and 43%, respectively, of the radiant energy to reach the body [137]. Concurrently, the dark apparel discharged 378 Wm^{-2} (73%) of the study participants' body heat to the ambient air, while the white garments only discharged 145 Wm^{-2} (28%). These results mean that in hot environments, dark clothing yields a better thermal comfort on the skin than does white clothing. Therefore, the wearer feels more comfortable in black than white clothing.

In general, the ability of a fabric to take up heat depends on its color (i.e., its *emissivity*), with black having the highest emissivity (100%). Assumed that you aren't sweating and that the fabric is at outside air temperature, the energy balance between your body and the clothes' fabric is given by the body-heat flux density and the fabric's emissivity and reflectivity. As a result, at same temperature, black fabric reflects the least, and absorbs the most heat. Of course, as you sweat, the moisture flux, and fabric's permeability for water and/or water vapor may strongly modify the energy balance, as discussed in Ch. 5. However, this flux doesn't alter the behavior of the radiant heat fluxes for white or black fabrics. Due to these reasons, wearing white between *Memorial Day* and *Labor Day* fails to enhance thermal comfort.

6.3 Why Do Americans Wear White in Summer?

So, what's the reason for Americans wearing white in summer? Indeed, the old fashion rule not to wear white before Memorial Day and after Labor Day[44] has rather historic, societal and environmental than thermal comfort reasons. In the East Coast cities of the 18th

---

[43] Tuareg is a tribe living in the Sahara Desert.

[44] Note that at that time, the two holidays were not yet established.

century, the urban heat-island effect combined with the muggy hot air and air pollution from the early industrialization reduced the quality of life in summer. Without AC and/or electric vans, any rooms heated up to uncomfortable conditions. Coal served to produce the energy needed for steam engines and other industrial processes. The sulfur dioxide, aerosols including soot released during the combustion process led to stinky, unhealthy air. Add the odor of horse pee and poops to the mix. These feces emitted volatile organic compounds (VOC), which may enhance or lower ozone concentrations depending on the concentrations of nitrogen oxides [138], [139]. Furthermore, in summer, the weather on the East Coast is often hot and humid between the frequent thunderstorms.

When you could travel back in time to the East Coast cities of the late 18th century, you would see pink skies, and a pink Sun on cloud-free days even around noon. The sunrays namely travel through a layer with many aerosols including soot. As a result, just the red light of the Sun's visible spectrum remained (Fig. 2.2). When a high-pressure system strengthened, visibility, and air quality worsened, the sky turned gradually more and more salmon-pink reddish (like during sunset). These red summer skies still occurred in the mid-19th century (see, e.g., Thomas Chambers' painting "Threatening Sky, Bay of New York").

In those days, hazardous smoke and muggy air conditions were a typical annoyance at the American East Coast in summer. In the cities, dirt was everywhere, not just from horse dung. Soot and dust fell out of the often-stagnant air, contributing further to dirty streets. Whoever was rich enough to retreat into a chalet in the mountains (Catskills, Adirondacks, Appalachian Mts., Vermont), at the Atlantic coast (Newport, Long Island, Martha's Vineyard, etc.) or on a lake would do so during summer.

Away from the cities, the air and environment were clean. The elevation made temperatures more bearable because temperature usually decreases with height in the atmosphere below 6.21 miles (10 km) height or so. The relatively cooler water of the Atlantic than the land can lead to *sea-breezes* and welcome cooling as well. At lakes, the water took longer to warm up than the adjacent land due to the higher heat capacity of water than soil (Tab. 3.1). Consequently, the cooler water than land surface reduced air temperature along the lakeside [4]. Due to the lower ambient air temperatures, rooms stayed at lower temperatures in these rural areas than in town. Consequently, summers in the countryside were pleasant.

In the cities, the dirt and working conditions forbid people like Jane and John Doe to wear white for practical and time reasons. Obviously, light-color clothes look dirty more easily than dark clothes, and require washing more often. However, at that time, people worked 12 h or more in a factory on six days a week, and had to wash their clothes by hand. On the contrary, the Rich had their personnel to wash, and iron clothes. Consequently, the Rich could afford to wear white - even when coming into or leaving the city.

Most likely, the meteorological conditions and cleanness of the countryside together led to the restriction of white to the warm season. Between what today is known as Memorial Day and Labor Day, the Sun's radiation reaching the surface is strong on cloud-free days (Fig. 2.1). Therefore, minor stains would bleach when laundry was drying outside. Furthermore, during summer, laundry dries faster than in winter due to the exponential relationship of the saturation water-vapor pressure and temperature (Fig. 4.1). At 90% relative humidity, for instance, warm air can take up more water vapor from wet clothing than cold air at same relative humidity.

These societal and physical reasons gave white clothing the association with luxury. Like always in history, people dreamed of wearing what the upper class wore. Therefore, the middle class (which only encompassed a few people at that time) started wearing white as *Sunday's Best* in summer. When working conditions improved in the 20th century, and cities became cleaner, more and more people picked up wearing white in summer because it seemed to scream "This person has money." Still today, many people consider dark and light colors more appropriate for cold and warm weather, respectively.

6.4 The Role of Dye for Comfort in the Touch

Historically, Levi Strauss, who introduced the jeans, made the first denim trousers cut exactly the same like jeans, but he dyed them brown. These brown pants were called *ducks*. However, the ducks became less popular over time because the blue version was so much more comfortable on the skin than its brown twin.

Coloring cotton denim in a color other than indigo involves sulfur dyeing. Sulfur dyes are non-soluble in water. At temperatures around

176F (80°C) and *alkaline*[45] pH-values, the dye particles disintegrate in the solution in the presence of a reducing agent (e.g., sodium sulfide, sodium hydrosulfide). Under these conditions, the dye particles become water-soluble, and the fabric can absorb the dye. Salt facilitates the absorption. After removing the fabric from the dye solution, the dye oxidizes in air. This oxidation makes the color insoluble in water when washing the garment later.

Hydrogen peroxide or sodium bromate are alternatives to create the oxidation in mildly acidic solutions. In contrast to indigo dye, sulfur dyes dye the thread thoroughly, i.e., also inside the yarn. Therefore, colored denim feels stiffer and less comfortable on the skin than indigo-dyed denim. Because indigo only sticks to the outside of the threads, washing causes the loss of some dye, thereby creating the soft, lived-in feeling over time.

6.5 Color and the Heat Loss at the Outer Layer of Clothing

When a person leaves the house, their clothes are initially at room temperature. Once outside, the clothes gain or lose heat to establish a new equilibrium with the air temperature. Obviously, the heat loss/gain increases non-linearly with the increasing difference between the initial temperature, $T_{initial}$, and the outside temperature (Fig. 6.1).

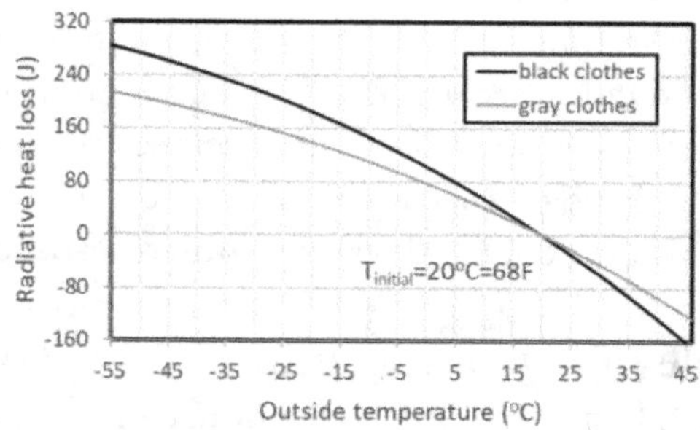

Fig. 6.1. Radiative heat loss of clothes initially at room temperature, $T_{initial}$, until the garment reaches outside temperature. An emissivity of 1 and 0.75 were assumed for black and gray, respectively. -45, -35, -15, 0, 15, 35, and 45°C are -49, -31, 5, 32, 59, 95, and 113F (cf. Fig. 1.5). Negative radiant heat loss values mean that the garment gains heat.

---

[45] While pH-values less than 7 are acidic, those greater than 7 are alkaline aka basic. A pH of 7 is neutral.

As expected from our experience, when the outside temperature exceeds the inside temperature, the outside of both the gray and black garments gains heat (Fig. 6.1). Because the black garment absorbs more solar radiation than the gray one, the outside of the black garment becomes warmer than the gray one. However, when it is colder outside than inside, the black garment loses more heat than the gray garment due to its higher emissivity. Because even layering only increases insulation for a limited time, it seems better to wear light than dark outerwear at temperatures below room temperature.

7 DRESSING FOR THE WEATHER IN THE EARTH'S CLIMATE REGIONS

7.1 Dressing for the Weather

Recall, the thermo-physiological and UV-properties differ with fabric structure (knit/weave type, tightness), thickness, weight, material or blends, color, and treatment. Furthermore, thermal stress or comfort depend on the weather conditions. As a result, the physical properties of the fabrics needed to mitigate thermal stress or to achieve thermo-physiological comfort differ among weather conditions (Tab. 7.1). Unfortunately, often a desirable fabric property may affect another also desired property negatively. Consequently, designing and building a weather-appropriate wardrobe often requires compromises. Recall, there are weather conditions for which one can't achieve thermal comfort for an undefined length of time. Instead, the clothing only prolongs the time of feeling comfortable.

Table 7.1 Ideal thermo-physiological properties for normal activity in various weather conditions. L, H, h, AP, R_t, UPF, OMMC, R_{et}, and WVTR are low, high, the fabric's thickness, air permeability, thermal resistance, UV-protection factor, overall moisture management capacity, and water-vapor transmission rate.

	h	AP	R_{ct}	UPF	OMMC	R_{et}	WVTR
Hot/warm & dry	L	H	L	H	H	L	H
Hot/w. & humid	L	H			H	L	H
Hot/w. & windy	L	L					
Cold & dry	H	L	H	H		H	
Cold & humid	H	L	H			H	

Recall, hemp is great for warm weather and active lifestyles because of its strength, durability, natural UV-protection, breathability, high absorptivity, and OMMC (Fig. 7.1c). It softens with each wash. Linen is highly absorbent and breathable and has excellent thermoregulation, making it suitable for hot and warm weather and active lifestyles. Its distinctive texture, durability, and natural drape add a touch of sophistication and timeless elegance.

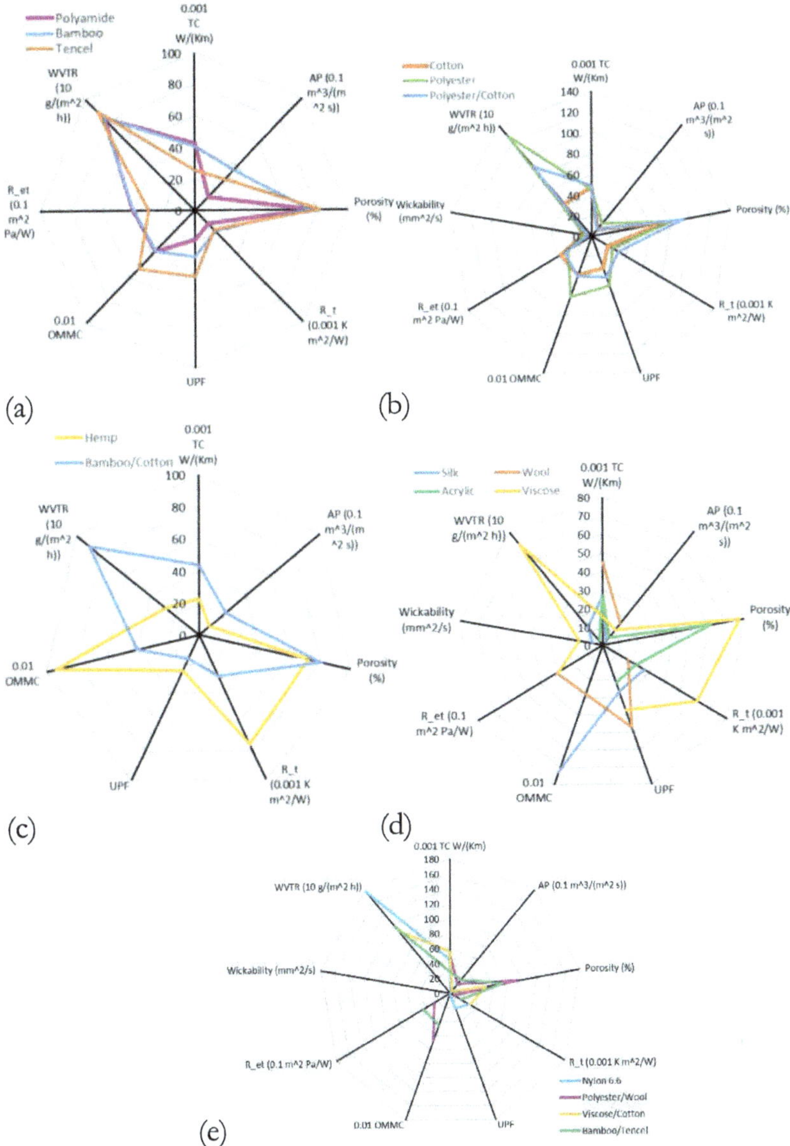

Fig. 7.1. Mean thermo-physiological and UV-transmission. Axes differ among panels. Open lines are missing data in (d) to (e). Data from [10], [14], [15], [20], [21], [23], [24], [25], [26], [27], [28], [29], [30], [31], [32], [33], [34], [35], [36], [37], [38], [39], [40], [41], [42], [43], [44], [45], [46], [47], [48], [50], [51], [52], [53], [54], [55], [56], [57], [58], [59], [61], [62], [63], [64], [70], [71], [72], [73], [74], [75], [76], [77], [78], [79], [80], [81], [83], [84], [85], [86], [87], [88], [89], [90], [91], [94], [95], [96], [97], [98], [100], [101], [102], [103], [108], [109], [110], [111].

Figure 7.1 shows radar plots of the mean UV-transmission properties, UPF, thermal conductivity, TD, air permeability, AP, thermal resistance, R_t, and water-vapor resistance, R_{et} for fabrics from various materials and blends. Herein the means are calculated over all data no matter of fabric structure, color, or treatment.

Typically, bamboo is naturally soft, breathable, highly moisture-wicking, good to very good in UV-protection (Fig. 7.1b), antibacterial, and hypoallergenic. Therefore, it is suitable for warm cloudy weather.

On average, Tencel, Modal, viscose, and bamboo-viscose offer softness, breathability, good moisture-wicking, good to very good UV-protection and medium thermal conductivity (Fig. 7.1a, d) making them suitable for warm weather. They all feel soft, and have an excellent drape. Polyamide has similar properties to bamboo except for a lower UPF and air permeability (Fig. 7.1a).

Wool fabrics like melton, loden, felt, duffle, are great for windy, rainy, humid, and cool weather down to slightly below the freezing point due to their high moisture regain, porosity, and low thermal conductivity. Thin wool fabrics are suitable for warm, hot weather as well. Cold and severe cold winters require layering and moisture protection.

In cases of blends, a generalization is not possible because thermal and moisture properties non-linearly depend on the blend ratio. Therefore, the mean values for blended fabrics (Fig. 7.1) have high standard deviations and differ notably in their higher statistical moments (not shown).

7.2 Solar Climate, Local Climate, and Weather

Dressing weather-appropriately requires checking the forecasts and looking out of the window. Creating or designing a wardrobe that is weather-appropriate for a certain climate region requires to understand the range of its typical weather and its local modification aka *local climate*. By no means, you can achieve thermal comfort by dressing for the climate means of a day due to the naturally high variability of weather.

Recall, the solar radiation at the top of the atmosphere depends on the time of day, day of the year, geographic latitude, and the angle of the Earth's rotation axis to the ecliptic (e.g., [16], [17]). These factors define the *solar climate* [140]. The Sun provides the energy that drives the general circulation of the ocean and atmosphere, the weather processes, and vegetation growth. Clouds, radiation absorbing gases and airborne particles reduce the solar radiation reaching the Earth's surface (Fig. 2.2) or your body.

Local climate depends on the solar radiation arriving at the surface, local geographic, water- and land-surface conditions. Oceans, lakes, land-cover and land-use influence the system *surface-atmosphere* by their albedo, emissivity, roughness, heat capacities, and the flux densities of matter (gases, particles), sensible and latent heat [4]. Due to the higher heat capacity of water than land (Tab. 3.1), land warms and cools faster than the water. As a result, water- and land-surface temperatures, and consequently, near-surface air and air pressure differ. Under light wind conditions (less than 13.42 mph, 6 m/s), these air- and pressure differences may lead to land- or sea-breeze systems in summer [4]. Furthermore, springs are cooler, while falls and winters are milder at the coast or near large lakes than at inland locations at same latitude and elevation [4]. This means despite the two locations have the same solar climate, their distance from the shore affects their local climate. Obviously, due to cold and warm ocean currents, the local climate also differs between the west- and east coasts of continents.

Shadow effects and the direction of mountain-slopes to the Sun cause differences in ground-surface and near-surface air heating. As a result, *mountain-valley-* or *slope-wind-systems* may form under light wind conditions [4]. In mountain valleys, during night, radiation loss can lead to temperature inversions. Cold air from the top of the mountains may move down the slopes and pushes the lighter, warmer air in the valley upward. As a result, the cold air is underneath the warm air. Because at same pressure, cold air is heavier than warm air, this state can last multiple days in high latitudes until a cyclone sweeps the cold air out. While the air is very stagnant under inversion conditions, narrow valleys are also prone for high wind speed, when they channel the wind. Therefore, people living in a small valley need both insulating and *wind-proof outerwear.*

Mountain barriers can lead to different local climates over short distance. Mountain slopes facing the major wind direction typically receive more precipitation than the slopes on the downwind side (*lee*) or the upwind flatland. Typically, the leeward side is drier and warmer than the upwind (*luv*) side because moist air loses moisture by condensation and precipitation while ascending. Because cloud-formation processes release heat to the ambient air, the humid air mass reaches the mountain top at a higher temperature than a dry air mass would. During the descend on the lee side, the air temperature increases by 17.6F (9.8°C) with every 0.621 mile (1 km) decrease in altitude. This local process is called *Föhn* in Europe, and *Chinook*[46] in North-America. Under such weather conditions, you need *waterproof gear* on the luv-side; but not on the lee-side.

Now, what is climate to begin with? According to the WMO, a climate mean is the average of the weather conditions over several (typically 30) years [141]. This means climate is a statistic on the weather elements like temperature, wind, relative humidity, precipitation, sunshine duration, etc. Of course, on a daily basis, the weather elements deviate from their 30-years daily average. Daily minimum and maximum temperatures[47] vary from their long-term means, too. Therefore, looking at the daily mean temperature and dressing for it is by no means sufficient for protection from frostbite, hypothermia, heat- or sunstroke. Add wind and/or high relative humidity to the mix (Fig. 5.3), and dressing for the weather becomes even more complex. However, looking at the local climate helps in creating/designing a wardrobe suitable for the local weather conditions.

7.3 Köppen-Geiger Climate Classification

In 1884, Köppen classified the thermal zones of the Earth according to the duration of hot, moderate, and cold periods and the impact on the organic world [142]. Therefore, these classifications are also referred to as *bio-climates*. In 1928, Köppen and his student Geiger classified the world's climate into six categories based on the mean annual and the seasonal behavior of monthly mean precipitation, and monthly mean temperatures [143]. Köppen updated the classification until his death. Thereafter, Geiger, other geographers, and meteorologists have updated the classification further (e.g., [144], [145], [146], [147], [148]).

The modified *Köppen-Geiger classification* (Fig. 7.2) represents the Earth's climate regions well even though it excludes temperature extremes, average cloud cover, solar radiation reaching the surface

---

[46] Chinook means "snow eater".

[47] The daily minimum and maximum temperatures are the lowest and highest temperatures of the day.

(e.g., number of days with sunshine), and wind. Because vegetation is a great climate indicator, we can consider this classification also as a good indicator for thermal comfort [18] and the clothing requirements.

Today, the modified Köppen-Geiger classification (Fig. 7.2) distinguishes twenty-five categories, each coded by three letters. The first letter gives the six major climate regions, tropical humid (A), dry (B), mild mid-latitude (C), severe mid-latitude (D), polar (E), and highland (H) climate.

The second letter subdivides the six major climate regions based on their precipitation. Here "w" represents dry winters. It means that in the driest winter month, the average precipitation is less than 1/10 of the average precipitation of the wettest summer month. Moreover, the summers (June, July, August in the Northern hemisphere; December, January, February in the Southern hemisphere) are dry. In the driest summer month, the mean precipitation is less than 1.2 in (30 mm), and less than 1/3 of the wettest winter month. The letter "f"[48] indicates moist climates with notable precipitation year-round and no further requirements on the precipitation conditions. Say welcome to the umbrella and rain-boots, or best start a collection. The letter "s" indicates dry summers. Sun-safe clothing is a *Must*.

The third letter of B-climates distinguishes between hot, "h", and cold, "k"[49], winters, while it measures the summer heat of the C- and D-climates. Herein "a" indicates that the mean temperature of the warmest month exceeds 72F (22.2°C), and that there are at least four months with monthly mean temperatures above 50F (10°C). The letter "b" means that the mean temperature of the warmest month is less than 72F (22.2°C), but there are at least four months with mean temperatures above 50F (10°C). The letter "c" translates into three or less months with mean temperatures above 50F (10°C). Concretely speaking, welcome your winter wardrobe. Finally, "d" means that there are three or less months with mean temperatures above 50F (10°C), and the coldest month has temperatures below -36F (−38°C).

Because in the Tropics, the solar radiation at the TOA remains relatively constant year-round (Fig. 2.2), the terms winter and summer make not much sense. Therefore, A-climates are only distinguished with respect to precipitation. Analogously, no real summer exists in E-climates.

---

[48] The letter "f" is the abbreviation for the German word "feucht" (moist).

[49] Hot and cold translate to *heiß* and *kühl* in German.

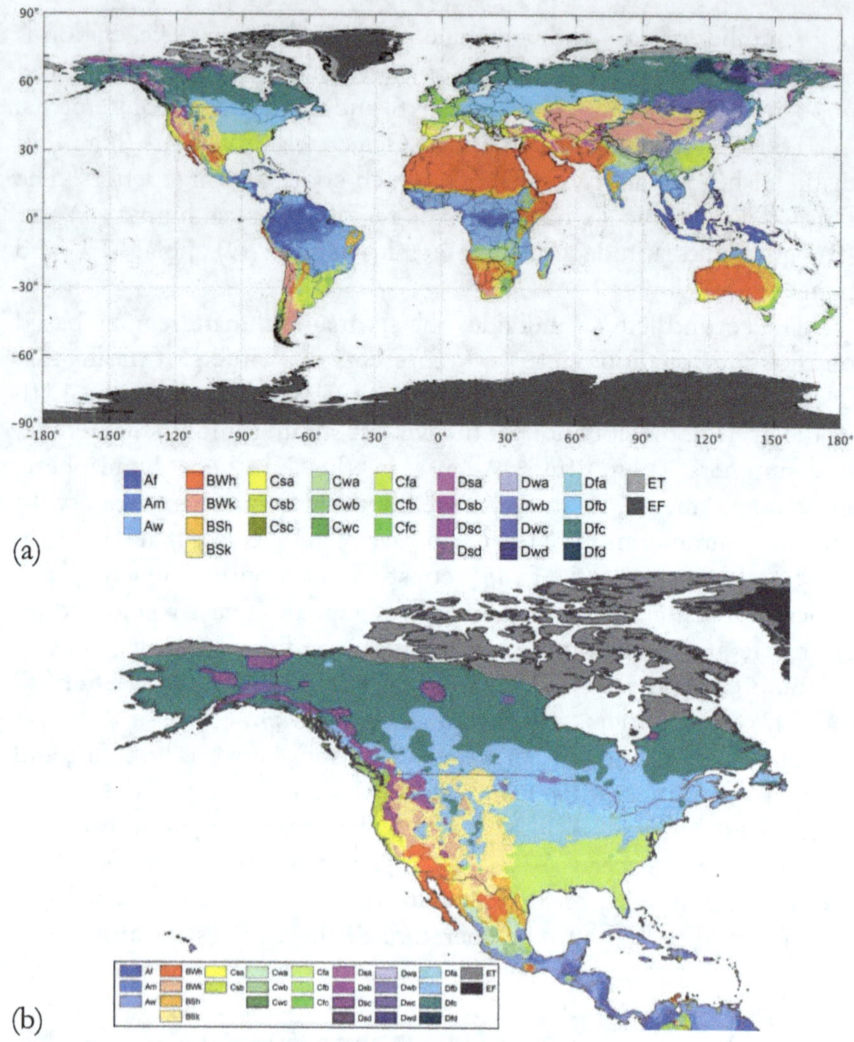

(a)

(b)

Fig. 7.2. Present-day (1980–2016) new Köppen-Geiger climate classification of (a) the World. Modified from [148]. The original is licensed under CC BY 4.0. (b) Köppen-Geiger classification for North America[50]. Cropped after [146]. Licensed under CC BY-SA 3.0. For explanation of the code see the text.

---

[50] A table of the Köppen-Geiger classification for US counties, boroughs, and cities and an interactive map exist at http://koeppen-geiger.vu-wien.ac.at/data/KoeppenGeiger.UScounty.txt, and https://www.plantmaps.com/koppen-climate-classification-map-united-states.php.

Obviously, the Köppen-Geiger climatology considers the solar radiation reaching the surface indirectly via the temperature and precipitation thresholds [17]. While wind is not included explicitly, wind direction plays a role in *monsoon climates*, and *moderate maritime mid-latitude climates*. In the latter, wind chill often occurs in winter. Of course, as discussed above, the local topography also affects wind speed and direction.

While there exist maps of monthly and seasonal wind climatology, they may be too coarse to really assess what you will face at a (new for you) location. Therefore, I highly recommend either contacting your local State Climatologist, regional climate center, or NWS office for information on the local wind climatology. When your region is quite windy, look for wind-proof outerwear (e.g., wind jacket, leather, shearling, tight knits/weaves depending on the temperature conditions).

7.4 Wear Comfort for the Weather in Tropical Climates (A)

A-climates exist at low latitudes, mostly between 15°S and 15°N (Fig. 7.2). Here, the *net solar radia*tion, which is the solar radiation reaching the ground minus the solar radiation reflected back to space, is large. The large net solar radiation and its low variability year-round lead to mean temperatures typically greater than 64F (18°C; Figs. 7.3-7.5). Day and night temperatures differ stronger than the mean temperatures of the coolest and warmest month. In many locations, seasonality occurs only in terms of wet and dry seasons.

The seasonal fluctuations of the *trade winds*, or *Asian monsoon* govern the weather of A-climates. Precipitation forms mainly along the *intertropical convergence zone* where the *north-east-* and *south-east trade winds* meet. Here, the warm, moist air is forced to rise, leading to cloud formation and precipitation.

In general, the weather conditions require lightweight fabrics with high breathability, and high air exchange like bamboo, cotton, hemp, linen, and rayon (Fig. 7.1). Linen is a good choice for professional attire, because the high humidity suppresses a bit the wrinkling of linen. Go for dark linen fabric because wrinkles are less visible in dark than in light fabrics. During high activity, high water-vapor permeability is needed to achieve cooling from perspiration. Windy weather conditions require light outerwear with low air permeability. Due to the rain, high water absorbency and moisture regain by the fabrics is key for wear comfort. Recall, at same percentage of excess moisture, low absorbent fabrics feel damper than those with high absorbency. Furthermore, wet fabric loses its insulation ability. Once the fabric starts to dry the related evaporative cooling might yield thermal discomfort, in the cooled-down air after the rain.

7.4.1 Wet Equatorial Climate (Af)

Af-climate occur mostly between 12°S and 12°N (Fig. 7.2). Typically, relative humidity is high leading to convective clouds, and strong cloud cover in the afternoon with precipitation in the late afternoon or evening. Annual precipitation totals amount 59-394 in (1500–10,000 mm; e.g., Fig. 7.3).

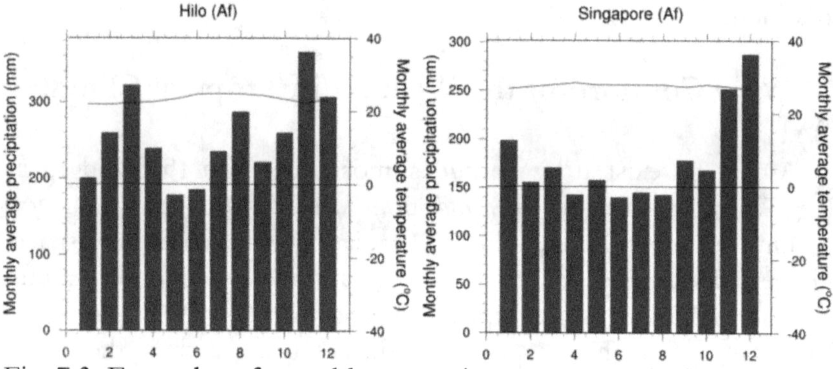

Fig. 7.3. Examples of monthly mean air temperature (red) and rainfall (blue) in Af-climate. 100 mm~3.9 in.

The weather conditions in Af regions mean frequent moderate thermal stress. Flip-flops, moisture-wicking fabrics, cool feeling fabrics, and umbrellas are weather-appropriate.

7.4.2 Tropical Wet-Dry Climate (Aw), Tropical Monsoon (Am) and Trade-Wind Climate (As)

Regions with *Tropical monsoon* or *Savannah climate* have a distinct wet and dry season (e.g., Figs. 7.4, 7.5), even though for different reasons.

During the rain-season, weather-appropriate dressing asks for flip flops or fast drying footwear, and raingear. The outer layer should keep the rain out, but permit high moisture transfer to the outside.

In Am climate, during the dry-season, the trade-winds may support the evaporation of sweat in the case of moisture-wicking fabrics. In both Am and Aw climates, wind-uptake of soil dust may stain clothing. Earth shades and blue look crisp and clean longer than light or dark colors. On days without rain, Modal, rayon, viscose, Tencel, and bamboo are good choices because of their moisture uptake. Satin weaves provide a cooling sensation. The dry-season and cloud-free days require sun-safe fabric.

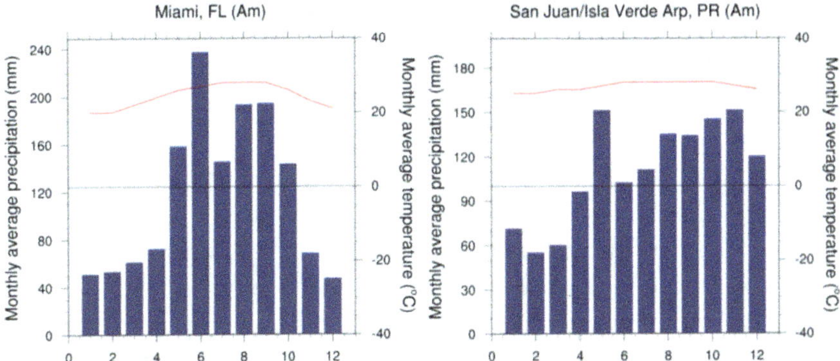

Fig. 7.4. Examples of monthly mean air temperature (red) and rainfall (blue) in Am-climate.

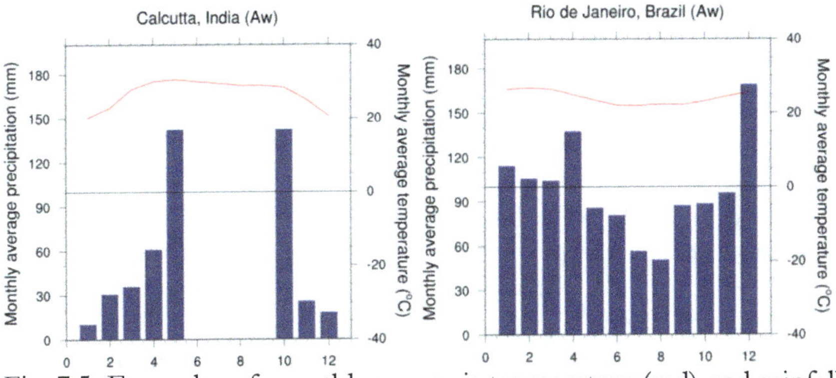

Fig. 7.5. Examples of monthly mean air temperature (red) and rainfall (blue) in Aw-climate.

The rare As-climate belongs to the wet-winter category. Typically, six to seven major storms occur per year, most often in late fall to early spring.

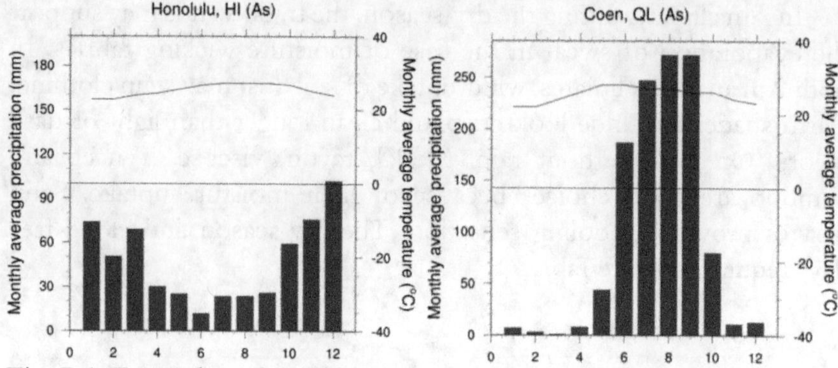

Fig. 7.6. Examples of monthly mean air temperature (red) and rainfall (blue) in As-climate. Coen is on the Southern hemisphere.

7.5 Weather and Wear Comfort in Dry Climates (B)

Arid and semi-arid climates cover about a quarter of Earth's land surface. They are mainly found in the 15–30° latitude belt of both hemispheres, but may reach up to 50° in some locations (Fig. 7.2). In this zonal belt, *semi-permanent high-pressure systems* dominate the weather. In high-pressure systems, air mainly sinks down, even though occasionally convective showers may occur. Consequently, B-climates exhibit low precipitation (Figs. 7.6-7.9), high year-to-year variability of precipitation, high evaporation rates (when water/sweat is available), low relative humidity, clear skies, and intense solar radiation (cf. Fig. 2.1). Consequently, wear sun-safe clothing year-round.

In hot, dry weather, the high intensity of solar irradiation requires to reduce the absorption of solar radiation. Therefore, fabrics should block the heat from the outside to enter your clothes, but let access body heat and water vapor from sweat pass. Low thermal conductive materials can reduce heat gain by conduction. Unfortunately, materials with low thermal conductivity can also block the sweat evaporation which requires maximized heat transfer.

In hot as well as in hot semi-arid desert climates, and in cool desert climate in summer, choose black, breathable, and moisture-wicking fabrics. In hot climates, people consider fabrics with thermal resistance lower than 0.05 m^2K/W as comfortable (e.g., cotton, cotton/flax, acrylic/cotton; cf. Fig. 3.2a). Because of their better moisture management performance, polyester/wool blended fabrics provide better wearing comfort than 100% wool fabrics (Fig. 7.1) [100]. Viscose and Tencel fabrics pose a small barrier to water-vapor transfer.

As a result, during sweating, they feel drier than cotton or cotton/Modal [149]. Tencel single-jersey, and polyester mesh-knit fabric or their combinations are great in hot weather as well. Cotton sweaters are great for winter at temperatures above 50F (10°C). Due to their low thermal conductivity coefficient, cotton or bamboo blends with Lycra are favorable for the cold season [74] in areas with warm or cool winters.

7.5.1 Bwh Hot Desert Climate and Bsh Hot Semi-Arid Climate

Areas with Bwh hot desert climate experience very little rain (e.g., Fig. 7.7), and have the highest summer temperatures on Earth.

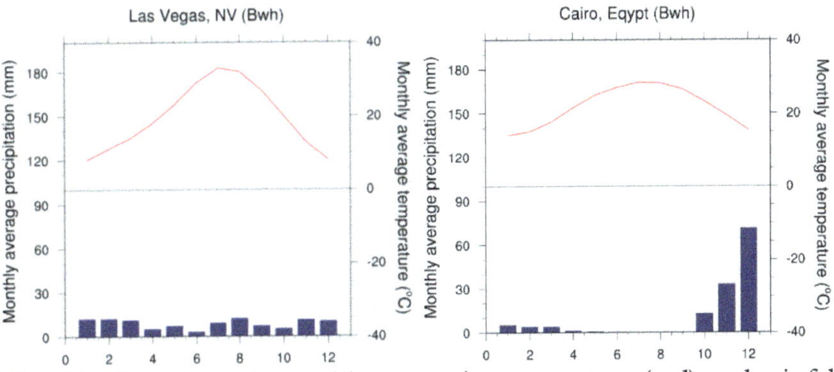

Fig. 7.7. Examples of monthly mean air temperature (red) and rainfall (blue) in Bwh-climate.

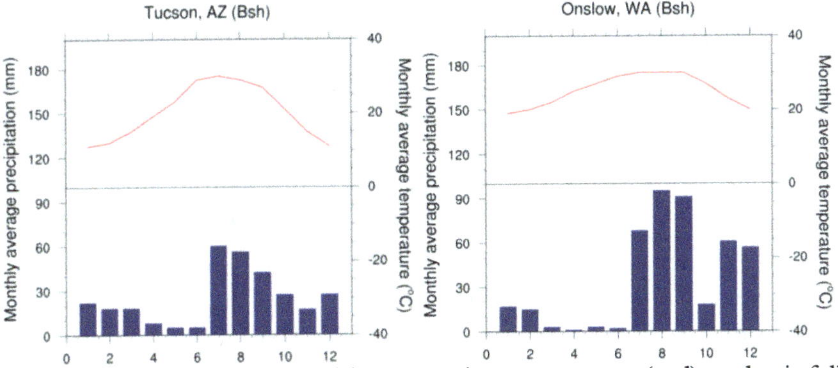

Fig. 7.8. Examples of monthly mean air temperature (red) and rainfall (blue) in Bsh-climate. Onslow is in Western Australia.

Due to the locations of Bsh close to the hot deserts, the rain occurring in hot semi-arid regions is marginal.

7.5.2 Bwk Cool Desert and Bsk Cool Semi-Arid Steppe Climate

Typically, cool deserts exist far away from the ocean or leeward of mountains (Fig. 7.2). This means these cool deserts are cut off from advection[51] of moist air and precipitation in form of rain or snow. In winter, temperatures often drop below the freezing point. Therefore, garments must provide appropriate insulation in cool steppe winter.

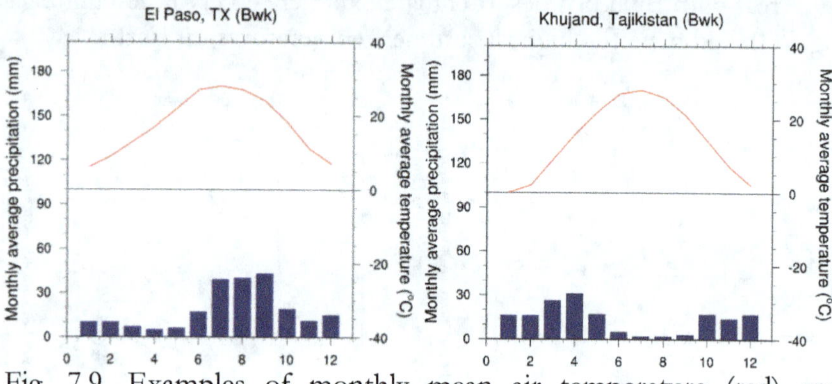

Fig. 7.9. Examples of monthly mean air temperature (red) and precipitation (blue) in Bwk-climate. Khujand was known as Leninabad.

While in Bsk-climate, the weather is similar to that in Bwk-climate, the former has more precipitation than the latter (cp., e.g., Figs. 7.9, 7.10). However, the total precipitation is insufficient for classifying the Bsk-climate regions as temperate or continental.

Note that the western central US (e.g., Montana, Wyoming, parts of southern Idaho, western Nebraska, western South and North Dakota, parts of Colorado) have thermal regimes that fit the Dfa-climate classification except for their total precipitation. Due to their dryness, these regions are generally classified as Bsk steppe climates (Fig. 7.2). Likewise, due to their semi-arid precipitation, the Canadian south-central and southwestern Prairie Provinces (e.g., Alberta) fit the Dfb criteria only from a thermal aspect. Therefore, these regions are also classified as Bsk. In both these regions, cotton is a bad choice in

---

[51] Advection is a meteorological term referring to the transport of an air mass from another region into the region under consideration.

winter. Therefore, when you live in one of these regions, read also the dressing advice for winter weather in Dfa- and Dfb-climate, respectively. However, go low-key on the humidity aspects.

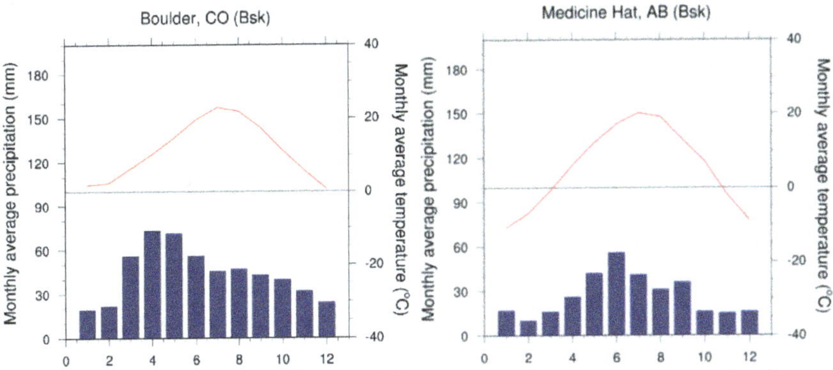

Fig. 7.10. Examples of monthly mean air temperature (red) and precipitation (blue) in Bsk-climate.

7.6 The Weather of Temperate C- and Continental D-Climates

The updated Köppen-Geiger classification identifies six C-climates and eight D-climates. C- and D-climates mainly occur in the mid-latitudes (mostly from 25° to 60°N and S), and high latitudes (north and south of 60°N and S), respectively (Fig. 7.2). In general, *mild mid-latitude climates* have an annual mean temperature above 50F (10°C), and monthly mean temperatures between 26.6F (−3°C) and 64F (18°C) in their coldest and warmest months, respectively[52].

In the mid-latitudes, day- and night temperatures differ less than the mean temperatures of the coolest and warmest month. Furthermore, the intensity of near-surface winds varies stronger in winter than summer. Typically, the *polar front* and related upper-level

---

[52] Some versions of the Köppen-Geiger classifications use 32F (0°C) as the lower temperature threshold for the boundary between the C- and D-climates. Choosing the freezing point to separate these climates locates the C-zone southward on the Northern Hemisphere, and puts northern Japan, northern China, and northern Korea into the cold D-group; in the USA, regions from the New York City metropolitan area including New Jersey and southern Connecticut, as well as the lower Ohio Valley, lower Midwest, and the southern Plains would fall in the mild C-group.

jet-stream govern the weather of the mid-latitude *Westerlies*[53]. The polar front and jet-stream are located at the northern (southern) and southern (northern) border of the mid-latitude Westerlies in the Northern (Southern) Hemisphere. The latitudinal and longitudinal location of the polar front and related jet-stream varies notably on the scale of several days to weeks, and with the season. During summer, both the polar front and jet stream shift poleward. As a result, *tropical air masses* can extend to higher latitudes of the respective summer hemisphere. On the contrary, the position moves equatorward in a hemisphere's winter. Consequently, tropical air retreats, and cold polar air masses move farther equatorward (cold-snap), and can even influence the weather in the subtropics. Typically, the occurrence frequency of *cold-air outbreaks* decreases with decreasing latitude; the frequency of subtropical air affecting the weather decreases with increasing latitude. Both these weather events go along with distinct changes in temperature and moisture conditions, especially, in winter.

7.7 Dressing for the Weather in C-Climates

In general, 50/50 bamboo/cotton, 50/50 soybean/cotton fabrics, low weft density 2/2 twill and matt twill fabrics are great for summer clothing in C-climates. While polyester/cotton blends with high cotton fraction, tight single-jersey cotton, bamboo and cotton/bamboo fabrics are excellent for summer, their loose versions are better for winter. Furthermore, 65/35 polyester/cotton fabrics are convenient for winter clothing, while 33/67 polyester/cotton yarns are suitable for summer clothing.

7.7.1 Humid Subtropical Climate (Cfa, Cwa)

Modal, rayon, viscose, Tencel, cotton, and bamboo are good choices for temperatures above 50F (10°C) under humid subtropical conditions. For a sense of coolness even on warm or muggy summer days, the body heat and evaporated sweat must be able to penetrate through the fabric away from your skin to the fabric's outer side. Obviously, thin fabrics are better for fast water-vapor penetration than thick fabrics of the same material. However, thin fabrics are more prone to transmit UV radiation (cf. Ch. 2).

---

[53] The term Westerlies refers to west wind (i.e., wind coming from west) being the dominant wind direction.

In Cfa-climate, precipitation shows no significant difference between seasons (e.g., Fig. 7.11).

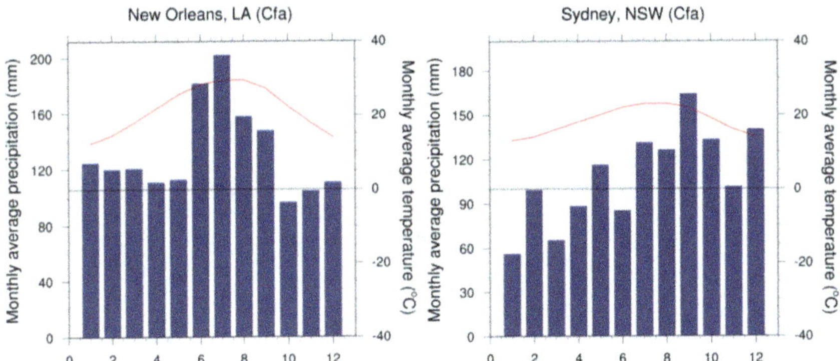

Fig. 7.11. Examples of monthly mean air temperature (red) and rainfall (blue) in Cfa-climate. Sydney, New South Wales is on the Southern Hemisphere.

On the contrary, Cwa refers to *Monsoon-influenced humid subtropical climate*. Here, the wettest summer month has at least ten times as much rain than the driest winter month (e.g., Fig. 7.12). Note that some authors define the precipitation characteristic of Cwa-climate as 70% or more of the mean annual precipitation occurs in the warmest six months.

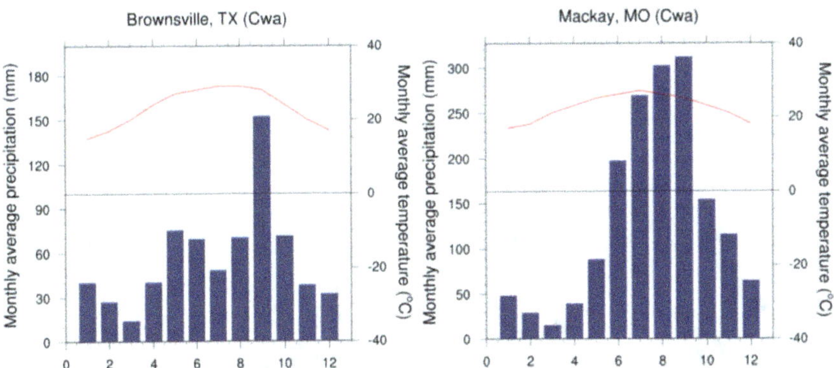

Fig. 7.12. Examples of monthly mean air temperature (red) and rainfall (blue) in Cwa-climate.

7.7.2 Mediterranean Climate (Csa, Csb)

In hot-summer Mediterranean Csa-climate, the wettest winter month has at least three times as much precipitation as the driest summer month, in which less than 1.6 in (40 mm) occur (e.g., Fig. 7.13).

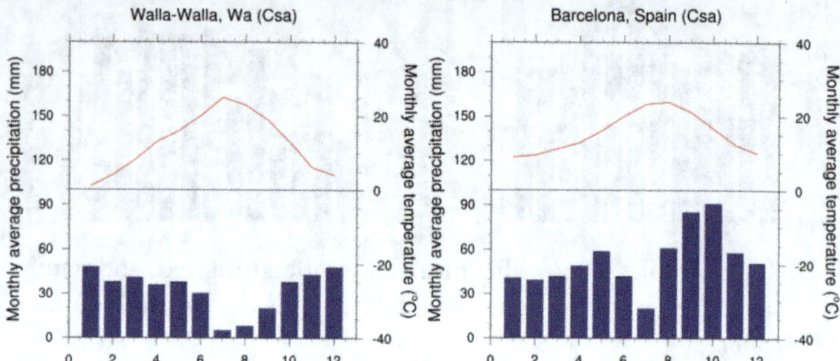

Fig. 7.13. Examples of monthly mean air temperature (red) and precipitation (blue) in Csa-climate.

Warm-summer Mediterranean Csb-climate has the same precipitation behavior than Csa-climate (cp. Figs. 7.13, 7.14). However, all monthly mean temperatures are below 71.6F (22°C) in Csb-climate (e.g., Fig. 7.14). Consequently, the weather of Csb-climate asks for thicker fabrics in winter than in Csa-climate. Lightweight, sun-safe, moisture-wicking fabrics are best in both climates in summer.

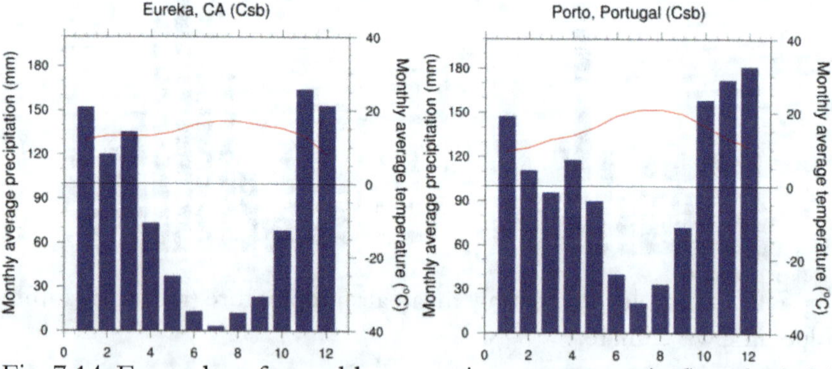

Fig. 7.14. Examples of monthly mean air temperature (red) and rainfall (blue) in Csb-climate.

7.7.3 Maritime Temperate (Cfb) and Maritime Subarctic Climate (Cfc)

People often refer to *maritime temperate* and *maritime subarctic climate* as *marine West Coast climate*. Because of the often-windy conditions, Cfc- and Cfb-climates can feel quite cold due to wind chill (Fig. 5.3). Wind chill and high relative humidity often lead to cold stress, especially, in winter.

Cfb-climate exists on the western sides of continents between 45°N and 55°N. In West Europe, Cfb-climate occurs in coastal areas up to 63°N in Norway (Fig. 7.2) and the coastal areas of Iceland due to the warm Gulf Stream. In Alaska, Cfb-climate exists along the Panhandle. Cfb-climates also occur at high elevation in subtropical and tropical regions (cf. Fig. 7.2). Most of New Zealand has Cfb-climate.

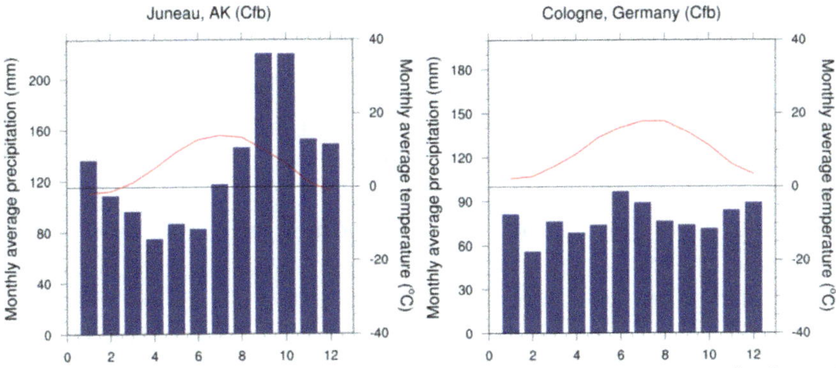

Fig. 7.15. Examples of monthly mean air temperature (red) and precipitation (blue) in Cfb-climate.

Because the polar front dominates the weather in Cfb-climate year-round, weather changes a lot with often overcast skies. At the US West Coast in Cfb-climate regions, spring and summers are cool (e.g., Fig. 7.15) due to the cool ocean currents. Generally, winters and falls are milder than in other climates at similar latitudes (cf., Fig. 7.2).

Maritime subarctic (Cfc) climate is also known as *maritime subpolar climate*. The name already sounds cold. This climate zone is poleward of the Cfb-climate zone (Fig. 7.2). Cfc-climate exists on narrow coastal strips on the western pole-ward edge of the continents (e.g., British Columbia, The Netherlands), and on islands off such coasts (e.g., Kodiak, AK, United Kingdom, New Zealand).

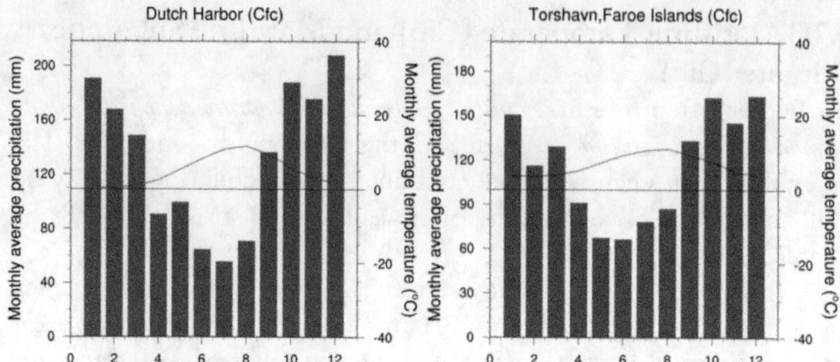

Fig. 7.16. Examples of monthly mean air temperature (red) and precipitation (blue) in Cfc-climate.

The winter weather in C-climates asks for layering. Any thin layers of air - like the air between triple windows in five-star houses - increase insulation. This being said you can achieve optimal thermal insulation with at least three layers (Fig. 3.21): A layer of underwear that is worn close to your body. Its purpose is to absorb sweat so it does not remain in contact with your skin. Ideally, an outer layer for protection from the cold air, wind, and potential rain is water-repellent, breathable, and insulating at the same time. It is important that the fabrics are permeable for water-vapor to let moisture pass from your body to the outside, but waterproof on the outside to protect you from rain.

However, there are trades-off to be aware of. Gore-Tex, plastic, titanium, and water-resistant coatings repel water. Down and wool-insulated coats provide good insulation and wind protection. The closer the weave, or knit, the less air is exchanged. Consequently, a loosely knitted wool sweater will let the wind through while the seemingly much thinner wind breaker or leather jacket won't. Because there is an optimum thickness for the insulation by thin air layers, the cut and fit of clothing matter. Skinny jeans, for instance, leave barely space for insulating air. As a result, body-heat loss is faster than for a slightly relaxed pair.

In between the outerwear and skin, still-air layers have to be created as additional insulation (Fig. 3.21). The goal is to prevent contact between your (tightly fitting) underwear, and your outerwear. This second layer serves primarily for insulation, but is also part of your indoor style. If the created layers are too thin, heat is exchanged too fast between your skin and the ambient air. To create an insulating air layer, this second layer has to be looser than the underwear layer, but not too loose that forced convection occurs (Fig. 3.1).

Ideally, at temperatures below 50F (10°C), your garments should reflect the thermal radiation of your body back to your skin. Minimize the heat transfer from your body to the environment with fabrics in a dense weave/tight knit (e.g., double-jersey, 1×1 rib, interlock), air-filled internal porous structures (e.g., flannel, lodden, melton, duffle, fleece) and/or low thermal conductivity. Wool blends with small fraction of cashmere are better than pure wool. Recall wool/cashmere blends keep you warmer than wool/acrylic blends, which are better than cotton/acrylic blends. The breathability of acrylic is far worse than that of wool. Wool blends, cashmere blends, and polyester blends are good choices due to their insulation and moisture deflection properties. Merino wool is excellent for humid conditions between 32-50F (0-10°C). Fabrics with low cotton content are convenient as well. For exercising in winter, spandex is the best sports material to keep you warm. Obviously, all other things being equal, thick clothes keep us warm longer than thin ones.

In winter, avoid getting damp clothing due to sweat, rain, or immersion because water conducts heat well (Ch. 3). Water, namely, replaces some or all of the air in the fabric's pores. Once the wet clothes start drying, evaporative cooling would add to the body-heat loss. Avoid cotton or other water-absorbing fabrics with low moisture regain at temperatures below 50F (10°C) under moist weather conditions. Instead, choose water-repellent materials. Wool, for instance, while being moisture absorbent, is water-repellent; it feels dry even when it absorbed moisture because the wetness goes into the fibers. Merino wool can regain 35% of its weight in water before it feels wet to the touch. Synthetics treated with wax, silicone, resins, or fluorocarbons[54] have similar water repellency like wool.

For the warm season, the most comfortable fabrics would be from yarns of cotton, bamboo/cotton or bamboo/viscose blends.

In Cfb- and Cfc-climates, winds can be chilly, and humidity is typically high. Consequently, wind- and waterproof coats are a *Must*. Because rain occurs year-round, get at least one rain-coat suitable for winter, and a lighter one for summer. Start a collection of umbrellas, and always carry a foldable umbrella in your bag.

---

[54] Fluorocarbons destroy ozone in the ozone layer.

Opt for shoes with rubber soles. Leather soles namely soak in the water from wet ground, and make your feet feel cold. If you absolutely are in love with that pair of shoes with leather sole, have a shoe-repair person glue a thick rubber sole onto the leather sole. However, be aware that the leather sole still can get soaking wet when you stay out in the rain too long or step into a puddle. Trust me, I have been there, felt that. Because these climates have a lot of rain, you may invest in a pair of rain-boots in a neutral and a fashion color for change and the weekend. Also, get a pair of snow boots for those snowy days.

Heavy-weight jeans, dress pants, dresses and skirts in wool or jersey fabric, and sweaters are in order in winter. Also, get some wool blazers to wear over your sweater or blouse/shirt on those chillier days. A wool pea coat and/or lined trench coat are good options for your way to work. On most winter days, pantyhose still does the job as demonstrated by Princess Katherine of Wales. When temperatures drop below the freezing point opaque black, gray, navy, nude, or brown tights work well with a sophisticated professional look while providing enough insulation. Like in all cold-winter climates, you need hats, scarves and gloves, and for the colder days a handful of long-johns.

7.8 Severe Mid-Latitude Climate (D)

Severe mid-latitude climates usually exist in the interiors of continents, or on their east coasts, north of 40°N (Fig. 7.2). Therefore, severe mid-latitude climate is subdivided into *humid continental* and *subarctic climate*. Severe mid-latitude climates have a mean temperature above 50F (10°C) in the warmest months. The coldest month has a mean temperature below 26.6F (−3°C). In the Southern Hemisphere, severe mid-latitude climates hardly exist due to the relatively small land mass between 40°S and 60°S (Fig. 7.2).

Most regions with severe mid-latitude climate are located underneath the upper-level, mid-latitude Westerlies throughout the year. The seasonal variations in the location and intensity of these winds and their associated features characterize the weather. Therefore, to assess whether or not you need wind-proof clothing look at the annual distribution of wind speed. That being said let me add the best cold-weather investments besides appropriate clothes are a remote cold-starter for your car, a thermal blanket to cover your car's radiator grill, and a motor-block heater to plug in your car at temperatures below -10F (-23.2°C).

Living in D-climates means your clothes have to close a large temperature gap (up to 100F; 37.8°C in Dfc-climate) from the summer heat to the winter freeze. Obviously, you have to learn to look at your wardrobe differently than you used to when living in a warm climate. American typical houses namely have similar size closets. Because your closet has the same size than that of a person living in sunny Florida, you cannot store a complete wardrobe for temperatures above 80F (26.7°C), between 60 to 80F (15.6 to 26.7°C), 40 to 60F (4.4 to 15.6°C), 20 to 40F (-6.7 to 4.4°C), 0 to 20F (-18 to -6.7°C), -20 to 0F (-28.9 to -18°C), -40 to -20F (-40 to -28.9°C) and below -40F (-40°C). Furthermore, you would not only overdraw your credit card, but also never find what you want to wear. Therefore, you have to wear your wardrobe in a broad way. It requires to have the basic essentials in your style in your neutral colors. Chose them in a way that you can combine everything with several (> 5) other pieces from your wardrobe. Also learn to incorporate items from the warm season in cold season outfits and vice versa. You have to shop like a fashion editor meaning that any item must work at least for three seasons. Furthermore, you need layering strategies to stay warm in winter, and have the right fabrics to stay cool in summer.

The clothing advices given in the following assume that you are not the kind of person who makes snow angles, and snow persons every day. Instead, they are for people whose cold weather activity is to walk from the house to the car or public transportation to go to work, do the groceries, pick up the kids at daycare, etc. and go back home. Of course, you need a down coat for the frigid cold days.

Recall, material properties, knit/weave density, and yarn treatment affect the insulation factor of your sweater. As a rule of thumb, tightly woven sweaters are best for windy regions. Cooked wool-felt is great in wet-cold climate. Qiviut keeps you warm for the longest. The style/cut determines whether or not you have cold bridges. Hat, scarf, and gloves are *Must-haves*.

In March, the Sun is still low in the sky. You need sunglasses during the day because of the refection of sunlight on snow and icy roads. The snow acts like a mirror for which you need sunscreen with high SPF for outside activities.

Because temperatures may go above the freezing point during the day in mid-spring, you need waterproof footgear with thick insulating soles. In footwear, insulation against conductive heat loss to the ground is most important in all seasons, but summer. Footwear for outside should be of minimal weight because 2.2 lb (1 kg) of footwear correspond to about 11 lb (5 kg) of torso load in the energy cost of walking. Avoid wearing ordinary plastic, rubber boots, or leather hiking boots in severe cold weather because of their high thermal conductivity (Tab. 3.1). Instead, wear boots fully lined with felt. On milder winter days, wool felt and shearling insoles work. Switch from your heavy boots to light dress shoes when at the office. Otherwise, your feet perspire. Once back outside, the sweat might become cold or even freeze.

When dressing for temperatures below freezing, the extremities are the real problem. You may want to use partial gloves that can cover the fingers with a mitten-type piece when you often have to use devices like a cell phone or laptop that require finger dexterity. Be aware that the relatively large surface area of fingers means a rapid heat loss because the heat exchange linearly depends on the surface area. Therefore, restrict your texting or wait until you are inside. After work on very cold days, wearing shearling mittens over your gloves when driving avoids getting cold fingertips from the frigid cold steering wheel. However, driving with mittens needs experience, which can be gained in an empty parking lot.

There is an Inuit saying, "When your feet are cold, cover your head." You can lose about 10% of your body heat through your head. As a result, in cold weather, cover your head with a hat or hood; whatever is practical and suits your style while providing enough protection. Your hat should be able to cover your ears, especially, when the wind is strong. A wind-proof hat like beanies or hats with ear-flaps can be far more convenient than a wind-proof hood due to the better freedom of movement and visibility. Stir away from pom-poms on top of hats. The pom-poms make a snug fit impossible, leaving a large air gap of less insulation. In windy climate, avoid dangly pom-poms on your hat or long scarves, as they may blow into your face and block your view. Even worse, they may remove your glasses or contact lenses.

In severe cold weather, double your socks/tights, but avoid that your footwear becomes too tight. Start with a thin pair as under-socks

(I like to use nude knee-socks or hosiery because of their professional look with pumps at the office.). Then, add a thick pair on top for good insulation – preferably in wool loop-stitch with small amounts of synthetics for stretch, snug fit, and durability. Avoid cotton socks or tights because their low insulation (Fig. 7.1) bears the risk for cold feet and frostbitten toes. Furthermore, cotton absorbs sweat and holds the moisture, resulting in a thin film of transpiration on your feet. Recall, once wet, cotton has zero insulation. On the contrary, socks/tights made of wool, IsoWool, or fleece and similar synthetics absorb excessive perspiration while keeping their insulation properties. Due to their larger thickness, they also can absorb more moisture than the cotton version. Moreover, these materials dry themselves out when being worn.

When you are into strenuous outdoor activities even at below freezing temperatures, invest in a few pairs of sock liners. A sock liner adds (marginal) extra insulation, and effectively transfers perspiration from the foot to the sock. Consequently, the sock liner itself, and the feet remain dry.

Your clothing must make sense from a practical point of view. For instance, a maxi skirt or dress is a good idea to protect your legs in cold-dry conditions when sidewalks are well-cleared and/or snow is so powdery that you can just shake it off. In wet-cold climate or where snow is slushy and/or thawed on the sidewalks, your skirt will be just a big mop and look like that once you reach your destination. Not a promotional bell ringer! A whimsy maxi dress or skirt are a bad idea in windy weather. The wind will stick it to everything in the skirt's reach. There is a high risk to rip your skirt into pieces. Furthermore, the wind will give you an (un)wanted workout.

7.8.1 High Altitude Mediterranean Climate (Dsa, Dsb, Dsc)

In North America, *Mediterranean climates* reach farther north than in Eurasia (Fig. 7.2). Dsa-, Dsb-, and Dsc-climate only exist at high altitude near regions with hot summer Mediterranean climate (Csa). Altitude increases from Dsa- to Dsc-climate. Examples of Dsa-climate are Spokane (WA), Erzincan (Turkey; Fig. 7.17), Lytton (British Columbia), and Saqqez (Iran).

Examples of Dsb-climates are the southern region of Lake Tahoe (CA) and Flagstaff (AZ) (e.g., Fig. 7.18).

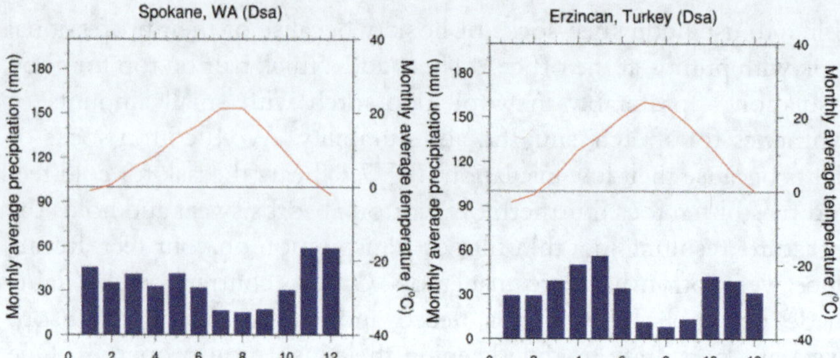

Fig. 7.17. Examples of monthly mean air temperature (red) and precipitation (blue) in Dsa-climate.

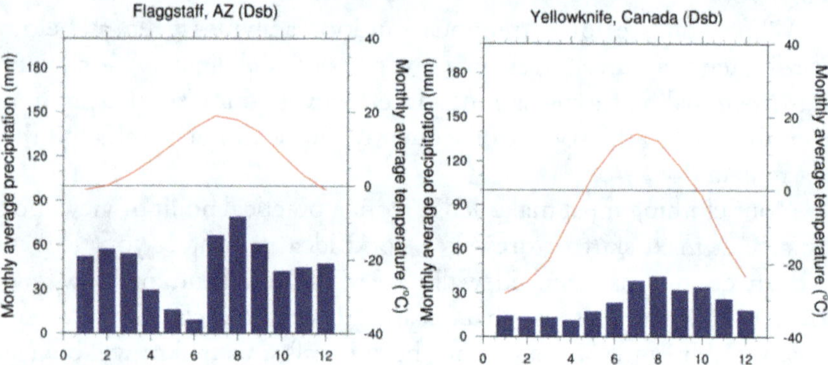

Fig. 7.18. Examples of monthly mean air temperature (red) and precipitation (blue) in Dsb-climate.

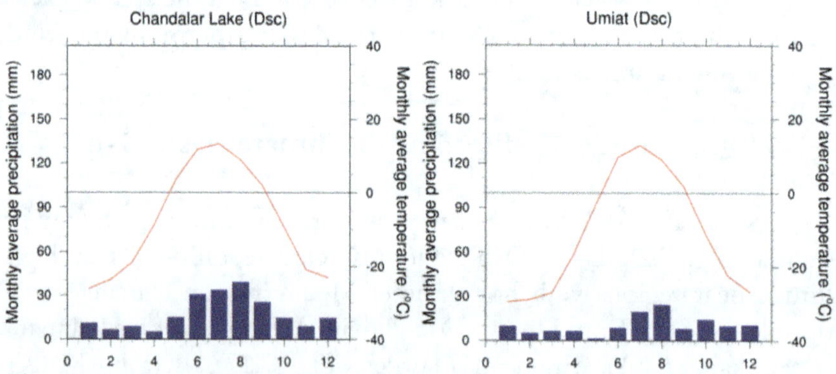

Fig. 7.19. Examples of monthly mean air temperature (red) and precipitation (blue) in Dsc-climate.

Dsc-climate translates into a pre-humid *Mediterranean snow climate* (e.g., Fig. 7.19). Examples are Galena Summit (ID), the upper part of the Massif Central (France), the Sierra Nevada (Spain), or Zubački Kabao (Montenegro).

7.8.2 Humid Continental Climate (Dfa, Dfb, Dwa, Dwb)

Humid continental climates occur over the continents between 30° and 60° (Fig. 7.2) in the major zone of the interaction between polar and tropical air masses. Therefore, these regions experience occasional incursions of Arctic or tropical air. Such incursions yield abrupt and strong temperature changes from one day to the next aka cold-spells or *warm-spells*[55]. In winter, temperatures may go up or down more than 20F (11.1°C) in 24 h. Such changeable weather occurs in all seasons and is a main climate characteristic. It is especially distinct in the eastern US and Canada, where hardly any notable topographic barriers exist to restrict the movement of air masses between low and high latitudes and vice versa[56].

Because of the possible drastic temperature changes, get rid of the bad habit to recycle yesterday's outfit today. Style-wise it is a faux pas to begin with. However, weather-wise, it may make the difference between feeling miserable or extending the time of feeling comfortable.

Severe cold winter weather requires layering as described for cold weather dressing in C-climates. However, you need more layers and/or thicker fabrics to enhance insulation. Style-wise you have to ensure that taking a layer off does not break your look. So, to speak, you have to style several outfits in one.

7.8.2.1 Dfa-Climate

The subcategory Dfa refers to *humid continental climate with severe winters*, no dry season, and hot summers. *Mid-latitude cyclones* (aka *extratropical cyclones*, or *frontal cyclones*) which form along the polar front, influence the weather in spring, fall, and winter. From fall to spring,

---

[55] Alaskans call weather situations under which tropical air reaches Alaska in winter "Pineapple Express".

[56] Milan, Italy (45.4640°N, 9.1916°E), for instance, has Cfa-climate like Houston, Texas (29.7631°N, 95.3631°W) despite Milan is located even farther north than Albany, NY (42.6525°N, 73.7567°W) that has Dfb-climate (Fig. 7.2). The presence of high topography (Alps) makes all the difference.

precipitation - often as snow - stems from these cyclones. A continuous snow cover exists for one to four months in many parts of this region, especially in the north. Snowfall often goes along with high wind speeds associated with an intense frontal cyclone and may occur in form of a *blizzard*. Clear, cold weather occurs when *continental polar air* governs the region.

Severe thunderstorms and even *tornadoes* may occur in early summer when the polar front is in the southern margin of the Dfa-climate region. Convective showers form when *maritime tropical air* moves northward behind a retreating polar front. Summer temperatures may reach in the 90s (>32°C). Due to the high humidity, the hot summers (Fig. 7.21) are often unpleasantly muggy.

Tokyo (Japan), Santaquin (UT), Boston (MA), Charlotte (NC), Des Moines (IA), Hartford (CT), and Toronto (Ontario) are examples of Dfa-climate.

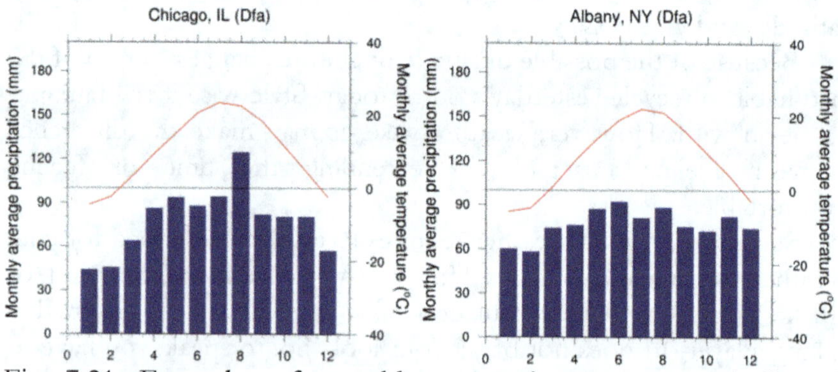

Fig. 7.21. Examples of monthly mean air temperature (red) and precipitation (blue) in Dfa-climate.

The weather of the Dfa-regions requires protection from precipitation year-round (e.g., Fig. 7.21). Your winter gear must be wind-proof. Boots must well close at your leg to hinder snow from entering your boots when you have to walk through deep fresh snow.

Recall, humid, cold, windy weather is very aggressive and can "chill you to the bone." Therefore, dress in layers (Fig. 3.21). The first layer should consist of quick-dry material in a snugly, but not tight fit. Long underwear made of polypropylene[57], silk or a thin fleece are best. The

---

[57] Note that when flying in helicopters or small aircrafts wearing polypropylene is prohibited due to its high inflammability.

intermediate layer(s) should be wool, fleece, polyester blends, or wool/cotton blends with low cotton content to insulate against the cold. The outermost layer should be water- and wind-proof. Look for good fasteners at the wrists and neck to keep moisture and wind out. You need a good, well insulating winter coat and a waterproof down coat knee-length. Trench coats with detachable (wool) lining are favorites among professionals. As wind can easily blow off hats, coats with a hood or hats that you can tie under the chin are best. Hoods protect better from frigid winds than a scarf plus hat. However, they impair your view because they partly cover and protect your face. Hoods also protect your hair from getting tangled, and avoid a hat-head.

In any windy climate, it is best to wear your scarf under your coat. Nevertheless, if you absolutely like the scarf-over-coat look, a short scarf hinders the wind to blow it into the faces of people passing by. Wide scarves are great to cover the lower part of your face in frigid-cold winds. Windy, cold weather requires wearing leather or otherwise wind-proof lined (cashmere, wool, or silk) gloves or mittens. In strong winds, protect your eyes with glasses. On bright, sunny days, sunglasses are a *Must* to protect your eyes from UV radiation and snow glare.

Comfortable boots are needed during snow and ice storms. Ankle boots work otherwise. Furthermore, you need at least one pair of well-insulating waterproof (snow) boots that hold up through snow and slush. Slip into some lighter dress shoes when inside the office. Treat your shoes with water-protectants to avoid damage from (melt) water and salt.

During summer, lightweight moisture-wicking fabrics like viscose/silk, cotton/polyester blends are key. Linen and cotton are great to beat the heat. Cotton seersucker avoids sticking to moist skin. Finding rest at night is particularly difficult in humid heat. Fabrics with cool sensation (all satin weaves), and high wicking capacity are best for sleepwear and bedding.

7.8.2.2 Dfb-Climate

Humid continental climate with severe winters, no dry season, and warm summers is categorized as Dfb (Fig. 7.22). Its summers are mild, while its long winters have frequent episodes of very cold conditions with clear skies. The weather varies immensely over the year, and the annual temperature range is larger than that of Dfa-climate (cp. Figs.

7.21, 7.22). Dfb-climate can occur with wetter summers than winters (e.g., Minsk, Belarus; Riga, Latavia; Tallinn, Estonia; Saskatoon, Saskatchewan; Calgary, Alberta; Winnipeg, Manitoba; Moscow, Krasnoyarsk, Nikolayevsk, Russia), drier summers than winters (e.g., Fargo, ND; Helsinki, Finland; Revelstoke, British Columbia) or uniform precipitation year-round (e.g., Buffalo, NY; Duluth, MN; Ottawa, Ontario; Halifax, Nova Scotia; Moncton, New Brunswick). Some isolated high-altitude locations in the western US like Aspen (CO) or parts of South Lake Tahoe (CA) have Dfb-climates.

Because Dfb-climate mainly differs from Dfa-climate by its summer conditions, (cp. Figs. 7.21, 7.22) your cold weather-protection strategy is the same as for Dfa-climate. However, your cold weather shoes should provide protection for lower temperatures than those in Dfa-climate. Despite the annual precipitation total being less than in Dfa- or Dwa-climates, you need protection from precipitation in form of rain and/or snow year-round.

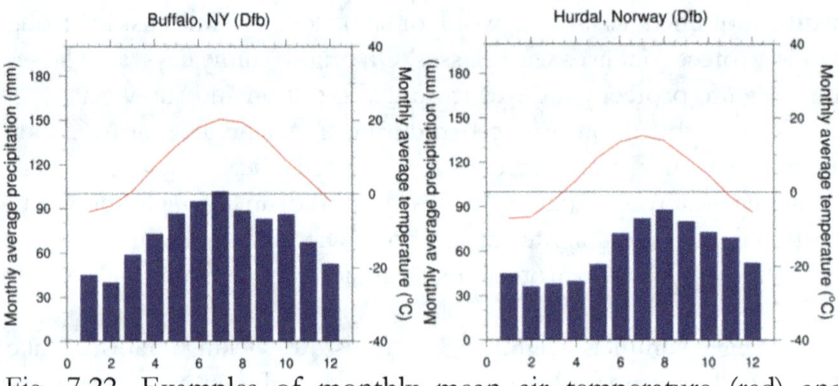

Fig. 7.22. Examples of monthly mean air temperature (red) and precipitation (blue) in Dfb-climate.

7.8.2.3 Dwa- and Dwb-Climates

Like in the Dfa-climate zone, mid-latitude cyclones affect the weather in spring, fall, and winter, and cold, clear weather dominates when *continental polar air masses* move into Dwa-regions. In summer, thunderstorms occur occasionally. Examples for Dwa-climate are Hyannis (NE; Fig. 7.23), Martin (SD), or Seoul (South Korea).

Like the Dfb-climate regions, Dwb-climate regions have mild summers and long winters with frequent cold-snaps under clear-sky conditions. The weather varies immensely in the annual course with a wide annual temperature range. The annual precipitation total is less than in Dfa- or Dwa-climates. Examples for Dwb-climate are Hill City (SD; Fig. 7.24), Vladivostok, Khabarovsk and Blagoveschensk (all Russia).

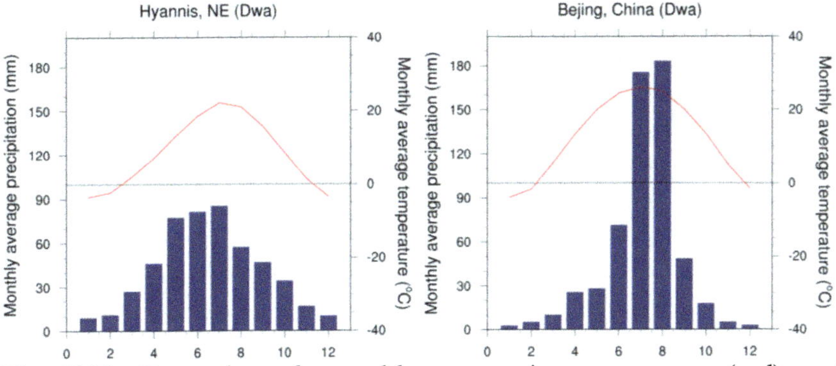

Fig. 7.23. Examples of monthly mean air temperature (red) and precipitation (blue) in Dwa-climate.

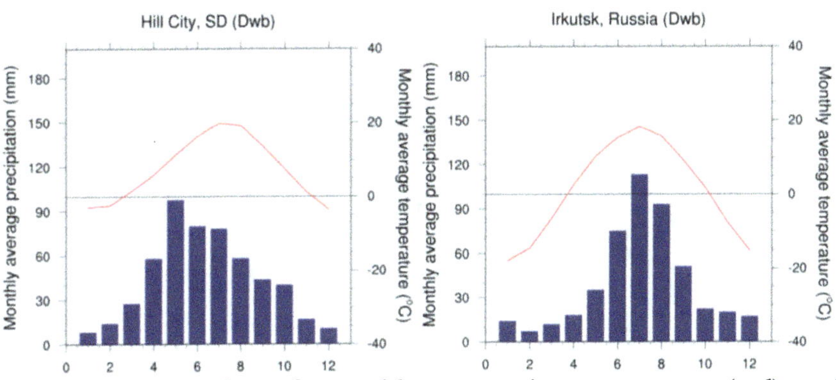

Fig. 7.24. Examples of monthly mean air temperature (red) and precipitation (blue) in Dwb-climate.

Except for the precipitation-protective gear, dressing in Dwa- and Dwb-regions requires the same strategy for protection from the cold as in Dfa- and Dfb-region, respectively. Humid continental climate with severe, dry winters and hot (Dwa) or warm summers (Dwb) does not need precipitation protection on a regular basis. Therefore, you only need one raincoat plus waterproof shoes or rain boots) for summer, and one waterproof winter outfit. For light rain in summer, you may consider an umbrella if it is not windy. The occasional snow requires non-slipping winter boots. Cleats work well on icy roads and sidewalks to protect you from slipping. However, they ruin your shoes! Therefore, wear them on shoes you do not particularly care about anymore, and switch to your beloved ones at the office. Wool-felt boots are great on dry days with temperatures between -18 and -8°C.

7.8.3 Continental Subarctic Climate (Dfc, Dfd, Dwc, Dwd)

Subarctic climate occurs between 50°N and 70°N from Alaska to Newfoundland and from northern Scandinavia to Siberia (Fig. 7.2). In *subarctic severe mid-latitude climate*, occasional mid-latitude cyclones influence the weather in spring, summer, and fall. However, continental polar and *Arctic air masses* dominate most time of the year. Long periods of bitterly cold weather with short, clear days, low humidity and relatively little precipitation characterize severe mid-latitude continental subarctic climate. This climate has short cool summers. Immense weather variability occurs over the year, and the annual temperature range is quite large (e.g., Figs. 7.25-7.28). Annual precipitation totals range between 10 and 20 in (250 and 500 mm). Many areas are underlain by permafrost[58] (Fig. 7.25).

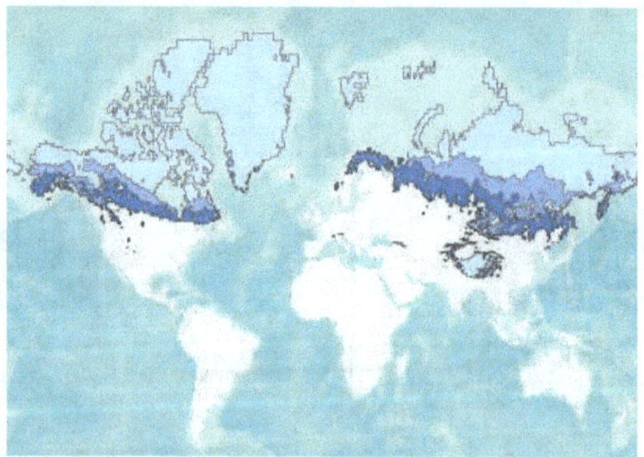

Fig. 7.25. Distribution of permafrost in the Northern Hemisphere. Continuous permafrost (light blue) means permafrost occurs everywhere. Discontinuous permafrost (medium blue) means the permafrost has scattered pockets of unfrozen soil. Sporadic/isolated permafrost (dark blue) means there are small, isolated islands of frozen soil in otherwise unfrozen soil. From [150]. Licensed CC BY 3.0.

---

[58] Any soil that remains frozen for at least two consecutive years is referred to as permafrost.

In general, in subarctic climate, you need heavy-weight clothing for winter, waterproof light to medium weight clothing for summer. Winter footwear asks for mukluks, onitsukas or when the snow is dry, lobben booties. Areas underlain by permafrost require waterproof boots with insulating sole for spring breakup. The permafrost namely inhibits the melt-water to infiltrate. Consequently, during snowmelt, water puddles are everywhere. Because in most regions, snowmelt is done after a week or two, one pair will be enough.

The Siberian or Canadian highs dominate the weather in the interior of the Eurasian and North American continents, respectively. In winter, subsiding air and long *dark days* lead to the formation of cold, dry air masses. As a result, the Dfc-climate region is the *source region* for continental polar and continental Arctic air masses. Once these air masses move south, they bring cold weather in the adjacent climate regions.

In summer, low pressure prevails over Siberia and northern Canada. Insolation is long, but not very strong. However, the daily totals are high (cf. Fig. 2.2). Humidity is not very high, but free convection, showers, and severe thunderstorms occur frequently.

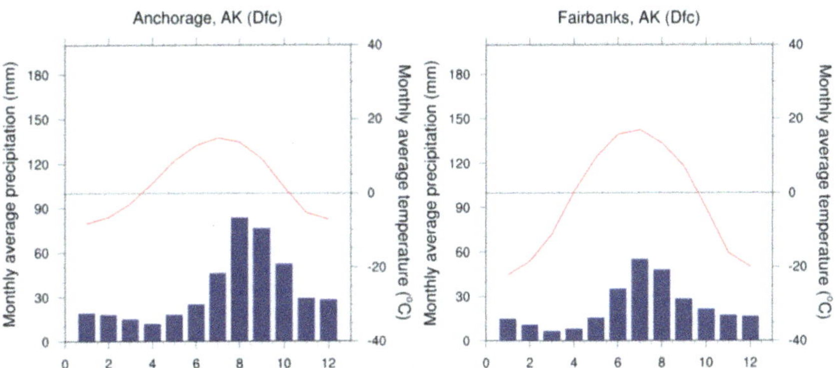

Fig. 7.25. Examples of monthly mean air temperature (red) and precipitation (blue) in Dfc-climates. Anchorage and Fairbanks have coastal- and inland Dfc-climate.

In the Dfc-regions, winters are severe, summers are cool, and there is no dry season. Dfc-climate occurs with both drier summers than winters (e.g., wide parts of British Columbia) and vice versa (e.g., Kirensk, Petropalovsk, Aliserkovo, all Russia; Schefferville, Quebec) as well as year-round uniform precipitation (e.g., Luleå, Sweden).

Examples of typical *coastal Dfc-climate* with wetter summers than winters are Cold Bay, Kotzebue, and Valdez (all AK; Fig. 7.25a), or Murmansk (Russia). Translated into clothing, coastal Dfc-climate requires rain-proof clothes.

Inland Dfc-climate can have summer maximum temperatures in the upper 80s (26.7-31.7°C). As a result, you can wear your shorts and tank tops beyond your annual vacation. Examples for Dfc-inland climate are large parts of eastern Interior Alaska, the Yukon and Northwest Territories or Moosonee (Ontario). While in the inland Dfc-region, winters are typically much colder than in the coastal Dfc-region, the former lacks the moisture (Fig. 7.25). Therefore, the inland Dfc-climate is felt much more pleasant for most people who lived in both coastal and inland Dfc-climate for a while. Inland Dfc-climate has the widest temperature range between its lowest and highest recorded temperature within the annual course. Consequently, when living in Dfc-climate you have to become a layering master to not run out of closet space. On extremely cold days, a moisture-proof down overall, thermo-pants or thermo-skirt worn over your normal clothing are options to stay warm longer. The skirts allow keeping the behind warm even when wearing a hip-length jacket.

When you live in a coastal Dfc-region, choose clothing with good fasteners at the wrists and neck to keep wind and moisture out.

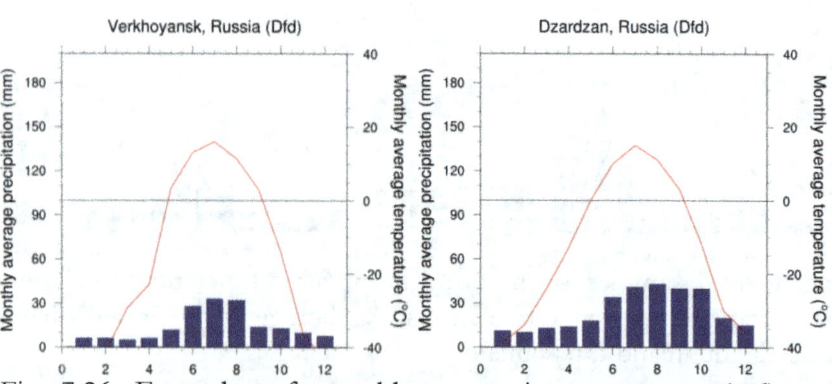

Fig. 7.26. Examples of monthly mean air temperature (red) and precipitation (blue) in Dfd-climate.

Category Dfd means severe, extremely cold winters, no dry season and cool summers. Oymyakon and Verkhoyansk, Siberia (Fig. 7.26) are examples of Dfd-climate. Because this climate is humid, see Dfc for how to protect from humidity and cold. Furthermore, recall that the letter "d" means that there are three or less months with average temperatures above 50F (10°C) and that the mean temperature of coldest month is below -36°F (−38°C). Therefore, you need at least a pair of cold weather boots (bunny boots, felt-boots, mukluks, or duck boots) suitable for temperatures below -40F (-40°C). For extremely cold days, wear a moisture-proof thermo-overall, thermo-pants or thermo-skirt over your normal layered outfit.

Category Dwc is the dry equivalent to the Dfc-climate (cp. Figs. 7.25, 7.28). This means winters are severe, and dry, and summers are cool. Examples for Dwc-climate are Tura and Bomnak (Russia).

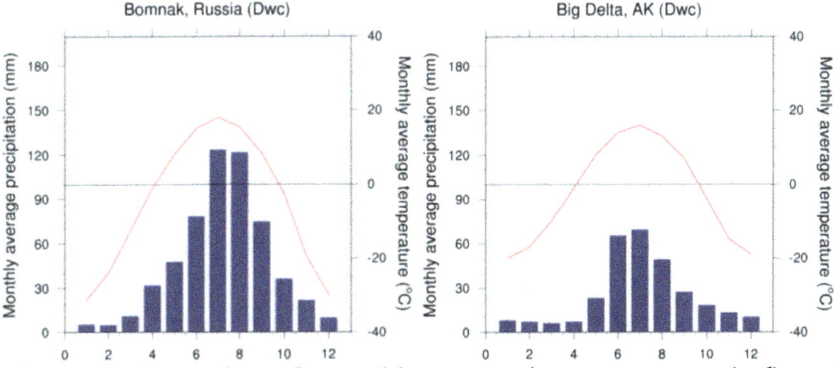

Fig. 7.27. Examples of monthly mean air temperature (red) and precipitation (blue) in Dwc-climate.

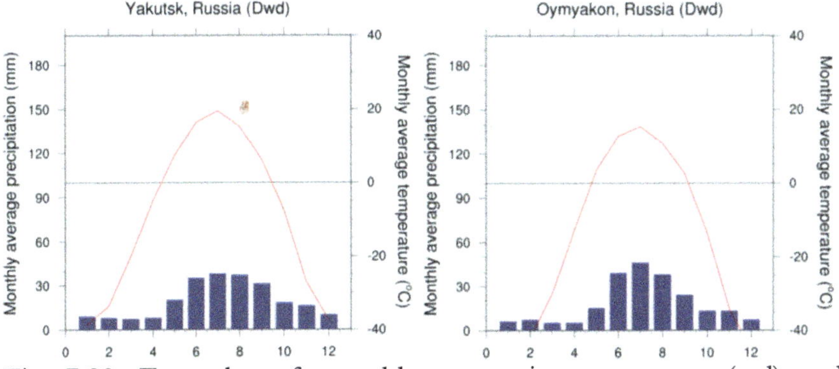

Fig. 7.28. Examples of monthly mean air temperature (red) and precipitation (blue) in Dwd-climate.

Despite during the severe, dry winters, extreme temperatures of Dwc-regions may be lower than those of cold, wet regions, you are lucky. Recall, it is much easier to protect from extremely dry, cold than cold, wet weather. Because Dwc is the dry equivalent to the Dfc-climate, read that section for clothing too, but focus less on the advice regarding humidity.

Category Dwd provides severe, very cold and dry winters, and cool summers (Fig. 7.29). Translated into fashion this means layering to create insulation in winter like in Dwc- and Dfc-climate.

7.9 Polar (E) and Highland (H) Climates

The Polar- and Arctic air masses of high latitudes (north of 60°N, south of 60°S) control the *E-climates*. These climates show a great diversity of subtypes, but all have in common low temperatures and precipitation (Fig. 7.30).

Updated Köppen-Geiger classifications consider the so-called H or *highland climate* (cf. Fig. 7.2). This class encompasses all highland areas that fail to fit in the other climate types.

The Köppen-Geiger classification for *polar climates* typically has no letters for seasonal precipitation. In these regions, falling and blowing snow are very difficult to distinguish. The classification distinguishes tundra (ET), and *snow-and-ice* aka *ice cap* (EF) climate.

In *tundra polar climate*, the mean temperature of the warmest month range between 32F (0°C) and 50F (10°C). In summer, the notable depletion of the sunbeams, the low sun angle, and the high reflectivity of snow keep the temperatures low despite the long daylight hours (white nights). Consequently, summer temperatures only slightly go above the freezing point on average. Welcome your winter wardrobe year-round, but start a nice collection of sunglasses to protect your eyes from the snow glare.

Because summer temperatures go slightly above freezing the upper part of the soil thaws. This layer - called the *active layer* - overlays the permafrost. The meltwater cannot infiltrate into the permafrost. Consequently, the upper soil may become very muddy, and puddles may form. Therefore, you will need waterproof shoes with some insulation for the cold summer months.

In winter, the dark days (Fig. 2.1) lead to strong radiation loss and low temperatures. The often-high wind-chill requires Arctic face masks and wind-proof clothing.

If you happened to move to a region with *polar tundra climate* you can kiss true summer (outside of your vacation, of course) good bye. You will have to dress for "winter" in all twelve months. The only difference is how much you have to layer.

Prominent examples for polar tundra climate are Utqiagvik (formerly Barrow, AK; Fig. 7.29), Alert and Iqaluit (both Nunavut), Nord (Greenland), Vardø (Norway). Non-arctic tundra climate occurs on Mt. Washington (NH), and Mt. Fuji (Japan).

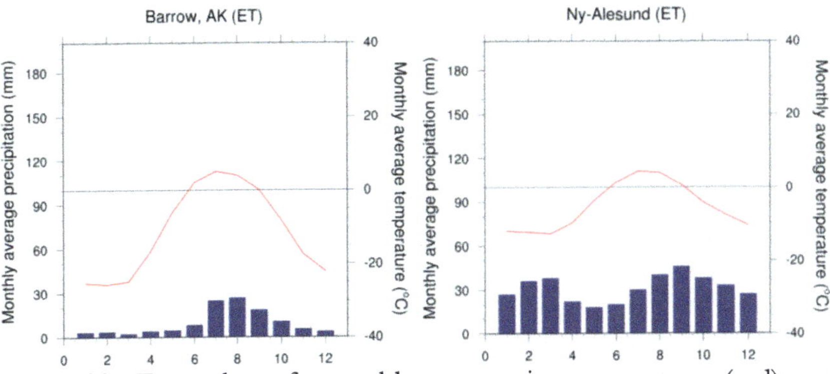

Fig. 7.30. Examples of monthly mean air temperature (red) and precipitation (blue) in ET-climate.

Polar ice-cap climate means perennial ice. It is the dominant climate of Antarctica and inner Greenland (e.g., Eismitte). Here, monthly mean temperatures remain below freezing year-round. If you move here, congratulations and welcome to being part of a polar expedition. No style allowed, no fashion fun, just wear what the expedition rules demand (Fig. 1.6), and look forward to more style at home.

However, there are some very small regions on Earth with non-scientific population living in EF-climate. These settlements are typically due to important mining activities like, for instance, La Rinconada, Peru. This city of 30,000 is located near a gold mine in the Andes, and is also the city at the highest altitude in the world. Therefore, one may argue it is an example of highland climate (H).

7.10 Cold and Hot Stress in Various Bio-Climates of the World

Given the UTCI stress categories (Ch. 5), thermal cold or heat stress cause different responses depending on the bio-climate region.

Cold-spells, for instance, can cause notable and often significant increases in mortality in people 65 years or older from cardiovascular causes, especially strokes, coronary heart disease events, and respiratory causes in subtropical Cfa- [3], mid-latitude Cfb- [1], and high latitude Dfb-climate [2]. In regions with warm winters, mortality from cardiovascular diseases increases with decreasing temperature; in Ireland (Cfb-climate), the cardiovascular diseases mortality rate is 45% higher in December than August, while it is 28% in the much colder winters of Norway [123] (~46% in Dfc-, ~26% in Cfb-, ~26% in Cfc-, ~1% in ET-, less than 1% in Cfc-climate).

Cold and heat stress of different severity can occur in all climate zones. Investigations in Europe [151] showed, for instance, very strong heat stress in Kołobrzeg, Warsaw, and Świeradów (all Poland), Budapest, Ljubljana (all Dfb), Milan (Cfa), Rome, and Athens (Csa) with maximum UTCI between 102.2 and 107.6F (39 and 42°C). In winter, Świeradów – a mountain site - and Warsaw had the lowest UTCI values due to local temperature inversions and advection of polar or continental Arctic air, respectively. The time with strong cold stress differed among sites, and was the longest (250 d) at Świeradów. Risk for cold stress existed on 210-220 d of the year in Warsaw and Kołobrzeg, and 160-180 d in Prague and Budapest. In Ljubljana, Rome, and Athens, cold stress occurred on 100, 70, and 120-130 d, respectively. In Dfb summer, the frequency of strong heat stress was less than 10%, while the frequency of strong cold stress was 40% in winter. The frequency for strong heat stress was 50-70% in Csa-climate.

Analyzing UTCI data from 1979-2017 for 456 sites revealed various degrees of cold stress for the bio-climates in Alaska (ET, EF, Dfb, Dfc, Dsb, Dsc, Cfc, Csb) with the highest frequency at the Alaska West Coast and along the Arctic Ocean [135]. In the continental and Inside Passage regions, the *no thermal stress* category occurred the most often. In Interior Alaska (Dfc), both strong heat and extreme cold stress occurred occasionally (Fig. 5.3). At most sites and in all bio-climates of Alaska, the absolute range between monthly means of daily minimum and maximum UTCI-values exceeded that of monthly means of daily minimum and maximum air temperatures. The major contributors to thermal discomfort (e.g., shortwave radiation, air temperature, moisture, wind speed) varied among bio-climate regions and in the diurnal and annual courses.

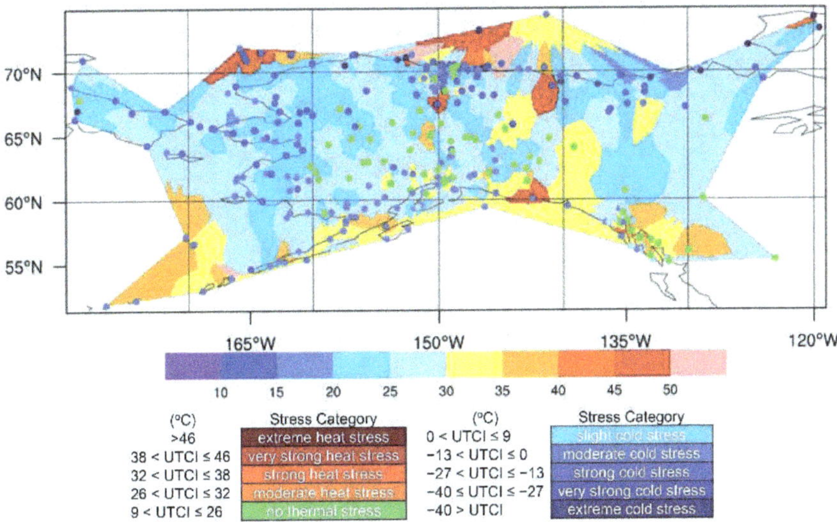

Fig. 7.31. Map showing the most often occurring thermal comfort conditions at each site (color-coded dots according to the stress categories) in Alaska and their percentage of occurrence between 1979 and 2017 (color-shaded areas, percentage given in the color bar) interpolated between sites. Note the sites in Canada and Russia served in the interpolation for improved coverage of Alaska. From [135]. .

Recall, T_{mrt} is one of the quantities used when determining the UTCI for identification of the thermal stress level. A study in Sweden showed that the mean radiant temperature, T_{mrt} can be much lower than the ambient temperature, T_a on extremely cold days with almost no solar radiation reaching the surface [128]. In Göteborg, Luleå (Dfc), and Stockholm (Dfb), the highest and lowest T_{mrt} occurred near sunlit walls and in shaded areas, respectively, during clear sky conditions. Under cloudy conditions, highest T_{mrt} occurred in open areas due to high solar diffuse radiation from the sky vault. Spatial variation of T_{mrt} decreased with increasing cloudiness.

While in winter, T_{mrt} strongly differs with latitude, no such sensitivity occurs in summer. In the Murmansk region, Russia, north of the Arctic Circle, thermal comfort varies significantly in space during contrast weather conditions [152]. While cold spells were more health-adverse to the inhabitants of Murmansk, Archangelsk, and Magadan than heat waves, the opposite was true in Yakutsk. During cold spells the mortality risks of people 65 years and older from ischemic heart disease, diseases of the circulatory system, and all non-accidental causes were 1.20, 1.14, and 1.12, respectively [153].

8 FUTURE DIRECTIONS IN TEXTILE ENGINEERING - PERSONAL COMFORT CLOTHES

8.1 New Demands for the Textile and Fashion Industry

As humankind developed more advanced skills, clothing also had to fulfill important ethical, and societal aspects (see Chs. 1, 6). Clothing became a way to communicate who you are, your societal status, group identity and wealth. Over time, fashion has even become a matter of political believes. Just think of the yellow or bluish ruff collars during Elizabeth's I reign, or the slogan T-shirt introduced by Republican Presidential Candidate Thomas E. Dewey in 1952. Today, feeling comfortable in your clothing also includes aesthetic, sensational, physiological, and psychological aspects. Special clothing is not only required for work, but also for play.

In the last decade, new markets opened up for the fashion industry. Many customers ask for *vegan fashion* either because they consider shearing of animals cruel or are concerned about the environmental impacts of overgrazing or tanning of hides; in case of fur, silk, and Canadian qiviut, the killing is the consumers' concern.

Other customers demand clothing in fabrics from "renewable resources". *Sustainability* became a selling point. Therefore, the textile industry started recycling fiber and filament from rotor spinning machines, and uneven yarns sorted out during quality control. Recovered fibers from rotor-spinning can be blended with raw material. A 20/80 blend of reclaimed and new fibers, for instance, has no notable difference in quality to 100% raw material. Also, cotton fabric with 15-25% waste fibers shows the same tenacity as fabric from only new fibers [154]. These and other recycling efforts improve the image of the textile industry, and reduce financial lost from waste, even though recycling also causes waste.

Hemp, linen, (virgin) bamboo, cotton, wool, cashmere and recycled polyester aka rPET are from sustainable resources. Recently, certifications such as Global Recycled Standard (GRS) and Recycled Claim Standard (RCS) ensure responsible recycling that meets specific environmental and social criteria.

Another trend in sustainability is upcycling of used clothes. Interestingly, we turn to what was the norm for the lowest class in Medieval times. It celebrates the imperfections and unique clothing. A paradise for creatives and designers.

Regenerated fibers like Tencel, Modal, bamboo-viscose, etc. are from plant-based fiber (eucalyptus tree, beech wood, bamboo, etc.) derived from wood pulp. Because the production requires (health-adverse) solvents, recent and future development will focus on closed-loop processing with recycling of solvents, and minimizing water and chemical waste.

Research on reducing water pollution from dyes, mordants, fabric treatments including bleaching and washing, etc. is another growing field.

8.2 Thermo-Physiological Wear Comfort: New Generation Yarns and Fabrics

Obviously, traditional fabrics like wool, silk, cotton as well as human-made nylon, and polyester have several shortcomings. Recall, cotton, for instance, fails to block the incoming infrared solar radiation in summer; in winter, cotton clothes require layering to increase insulation. Unfortunately, layering cotton results in bulky looks, and poor thermal management. Another drawback of most traditional fabrics is their lack of emitting thermal radiation from the body. In other words, they are unable to control the thermal emissions by the wearer's body [121].

Progress in medical research have increased the demand for advanced radiative textiles. Material science research has looked at conductive textiles, moisture regulative textiles, responsive textiles, air-cooled textiles, and active warming textiles (e.g., [121]). Recent progress in energy material- and nanofabrication technologies has permitted the creation of new materials for thermo-physiological comfort management. Sport enthusiast and people working in extreme climatic conditions will welcome these inventions.

8.2.1 Moisture Transference System Technology

Scientific research emphasizes that fiber type, yarn properties, fabric structure, finishing, and the cut of clothing affect a person's thermo-physiological comfort. Such research led to the development of the Moisture Transference System (MST) technology fabric that is very moisture-wicking. The MST-technology provides two benefits: You don't feel bathed in sweat, and the area for evaporation increases (your skin plus the fabric), thereby providing more cooling.

8.2.2 Functional Knitted Structures for Improved Wear Comfort

In the last decade, clothing engineering has developed so-called *functional knitted structures* to attain and maximize the comfort of clothing (Fig. 8.1). These fabrics address that atmospheric conditions might prohibit that sweat absorbed by conventional fabrics evaporates instantaneously. Therefore, to prevent a heavily sweating wearer feels wet/clammy engineers created a functional knitted structure that stores sweat. This fabric consists of two different layers. The separation layer is a hydrophobic textile material and builds the bottom side of the fabric that is in contact with the skin. The outside layer is an absorption layer consisting of hydrophilic textile material.

The behavior of functional knit strongly depends on the percentage of the hydrophobic component. Obviously, the linear density of the yarn plays no critical role. Instead, the number of *suction channels*, i.e., the porosity at the hydrophobic-hydrophilic material interface, significantly affects the thermo-physiological comfort achieved with structured knitted fabric [71]. The number of suction channels namely shows a quadratic relationship with both thermal absorptivity and thermal resistance.

Bi-layer knitted structures with polyester yarn as an inner layer and Modal/bamboo yarn as an outer layer show increasing water-vapor permeability with decreasing thickness, and openness in the fabric [155]. The moisture absorbency of such bi-layer knitted structures increases with increasing stitch density, and tightness.

Using polypropylene and cotton as bi-layer knitted structures maximizes the suction and moisture-transfer properties. However, again, the knit structure influences these properties. Open-construction 3D-eyelet, for instance, has better water-vapor permeability than micromesh, piqué, and mock-rib structures. Like for single-layer fabrics, thermal insulation increases with decreasing fabric density. The major factors governing the heat transfer through this bi-layer fabric remain fabric thickness, enclosed still air, and external air movement (wind).

8.2.3 Thermo-Physiological Comfort of the New Yarn Generation

In recent years, textile engineers have developed various yarns for activity-required properties. *Tetra-channel polyester*, for instance, wicks sweat away from your skin to the outer fabric layer. Due to the large surface area of the tetra-channel fiber, sweat evaporates rapidly. Because tetra-channel polyester fabric is highly breathable (Fig. 8.1) and requires low care, it is widely used for apparel, sports accessories, medical wraps, braces, and pads.

Another *new generation yarn* is a polyester fiber with slots that create a siphon effect. Consequently, fabric from this yarn pulls sweat away from the skin to the outer layers immediately. As a result, the fabric not only keeps you dry but also actively supports the regulation of your body temperature. No wonder that global top-apparel manufactures widely adopted this type of yarn for the fabrics of their casual wear, hosiery, lingerie, sportswear, swimwear, etc.

Another new generation yarn is a *patented blend of natural and synthetic* fibers. Herein, the natural fibers absorb moisture, pulling the sweat off the skin into the fabric. Synthetic fibers then repel the moisture, which forces the sweat through the fabric's surface, where it evaporates fast due to the ambient airflow. Consequently, the wearer stays dry, comfortable, and odor-free. In contrast to other topically-finished performance polyesters, this new yarn retains its wicking ability after washing (cf. Ch. 3). In other words, its comfort qualities are permanent.

Obviously, air permeability increases for fabrics knitted from these new generation yarns in single-jersey structure when the knit becomes looser. The fabric from the patented blend of natural and synthetic fibers is the most porous of the three types of new generation yarns [71]. Fabrics from channeled fibers provide better water-vapor permeability than traditional fabrics (cp. Figs. 3.9a, 8.1) because the channels provide a transport system. Because slot- and tetra-channel fabrics are both made from engineered polyester fibers, their thermal comfort properties are similar (Fig. 8.1). Fabric from the blend of natural and synthetic fibers creates better thermal insulation, and a warmer feeling at first touch than those form the polyester fibers. The reason is the lower thermal conductivity of cotton- than polyester fiber.

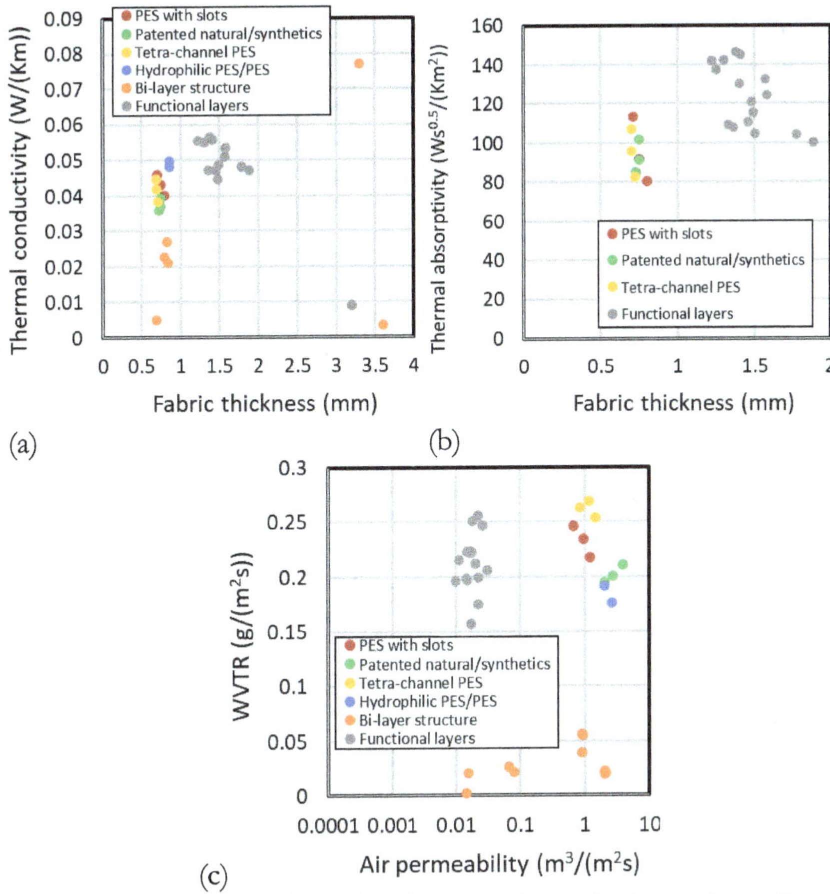

Fig. 8.1 Thermal conductivity (a) and thermal absorptivity (b) of functional structured knits and fabrics from the new yarn generation as dependent on fabric thickness. (c) Relation between air permeability and WVTR. PES is polyester. Data from [26], [71], [75], [78], [79]."

8.2.4 Self-Adaptive Materials for Improved Wear Comfort

Recently, self-adaptive materials have been designed to keep our body cool or warm and adapt to extreme weather conditions. These lightweight, breathable, flexible, and temperature-adaptive fabrics are efficient under both indoor and outdoor conditions. They can improve the thermal comfort of athletes, construction workers, firefighters, law-enforcement officers, and soldiers, as well as any individuals exposed to direct sunlight, and/or harsh weather conditions. Some of these advanced fabrics even serve to stabilize the health condition of patients with health disorders like the common cold, flu, chills, and neurological issues [121].

These innovations, among others, include fabrics that increase the thermal conduction and convection from the body to the ambient air to keep the wearer's body cool under hot conditions. Some cooling materials create radiative cooling by reflecting a maximal amount of the incoming solar radiation, while emitting heat from the body and fabric in the 800–1400 nm range to space [121]. Recall this range corresponds to the radiation at skin temperature.

Nano-porous polyethylene-based fabrics exhibit excellent radiative cooling because they are IR-transparent. Concurrently, they have sufficient water-wicking capability, excellent wearability, and good mechanical strength. According to studies, these fabrics can decrease skin temperature by 4.9F (2.7°C). Nano-porous polyethylene microfibers have similar softness as cotton. However, their nanopores scatter visible light. Consequently, fabrics from these fibers appear opaque in the visible range, while transmitting body heat in the mid-infrared range. As a result, nano-porous polyethylene microfiber fabrics keep the skin temperature up to 4.1F (2.3°C) lower than cotton fabrics of same thickness. Moreover, these new fabrics have better durability, and wearability than cotton fabrics [121]. Unfortunately, this control of the body's mid-infrared thermal emissions is only suitable for indoor conditions.

Outdoors, however, avoid solar radiation for daytime radiative sky cooling. As illustrated in Ch. 2, 93.4% of the solar irradiance occur in the visible light spectrum (400–700 nm), and near-infrared light (700–2500 nm). Consequently, a fabric designed for avoiding heat gain from sunlight must block the visible and near-infrared radiation. To achieve this goal, reflective materials like organic, inorganic, and natural compounds, and transition metals are introduced into fabrics. Other paths of achieving high solar reflectivity are 1) coating the fabric with calcined Titanium dioxide (TiO_2), 2) Chitosan-Titanium dioxide bi-layer and triple layer designs, 3) photonic structures, 4) impregnating with zinc-oxide nanoparticles, just to mention a few options.

Obviously, under cold weather conditions, your body temperature should be higher than the ambient air temperature. Therefore, personal clothing should minimize the heat transfer from the skin to the surroundings. To achieve this goal, lightweight materials with air-filled internal porous structure and/or low thermal conductivity can be used.

As a result, the garment retains the heat discharged from the body inside the narrow air layer between itself and the skin. To overcome the heat loss by thermal convection, the airflow within the layer and through the fabric should be minimized. A dense, woven layer can reduce the convective heat losses significantly. Highly reflective metallic fibers on the inside fabric layer of outerwear can reduce the radiative heat loss from the body to the outer layer and/or atmosphere. In such a three-layer outfit (Fig. 3.21), one can feel warm and comfortable for an extended amount of time.

What do self-adaptive fabrics mean for you, the wearer? First, it pulls down the myth that wearing white in summer keeps you cooler than any other color. Second, there exist fabrics that can actively cool you. Third, fabrics from natural sources and/or sustainable resources are not necessarily better than human-made fabrics when it comes to your thermal comfort. Fourth, there are cold and hot weather conditions under which thermal stress is inevitable.

8.3 The Surprise of Hollow Polyester

Sometimes developments of new fibers for improved thermo-physiological comfort may lead to unintended results. Nevertheless, they may be valuable for other purposes. Polyester fibers, for instance, can be created in various shapes of cross-section, including hollow. A study with round, round hollow, and trilobal fibers in twill- and plain weave showed increased thermal conductivity for fabrics woven with hollow fibers as compared with fabrics woven with solid fibers. Consequently, fabrics with yarns from hollow polyester fibers provided less insulation than those with solid fibers. This result contradicts findings found for naturally hollow fibers (e.g., angora), and air pores being good insulators (cf. Ch. 3). The reason is the larger outer dimension of the hollow than solid fibers at same fiber count. Consequently, the cover factor and total porosity of the woven fabric exceed the effect of inter-fiber pores [103]. This means the intended increase of insulation by introducing pores at the inter-fiber scale failed. But it may be useful when low insulation is desired.

8.4 Future Directions Related to Textiles

Given the increasing world population, more people will live in megacities which bear thermal stress from the urban heat island effect in summer. These local weather conditions require the development of fabrics with anti-bacterial properties to avoid odor in crowded public transportation, and allergic reactions. Because various blood sucking insects can cause infections and diseases[59], there is a need for insect-bite-safe fabrics as well.

The increase in the world population means also an increasing demand for clothing and fibers. Technology has to be developed to harvest the fibers from banana pseudo-stems in a cost- and time efficient way. Using the pseudo-stems for textile purposes instead of burning them as waste would also improve air quality, and provide a new economic branch in many tropical countries.

8.5 Final Thoughts on Textiles and Weather

In summary, textile engineering is an interesting field at the intersection fashion and STEM that is often overlooked by all genders when thinking about their choices of higher education. Moreover, studying atmospheric sciences has far more aspects to it than just climate and its changes.

---

[59] In medieval times, for instance, plague-infected rat fleas transmitted the bacterium, Yersinia pestis, responsible for the Black Death. Today, mosquitoes transmitting the West Nile virus - a single-stranded RNA virus - that causes West Nile fever are of big concern as are ticks. In the United States, tick-borne caused diseases encompass anaplasmosis, babesiosis, ehrlichiosis, Lyme disease, Rocky Mountain Spotted Fever, Southern Tick-Associated Rash Illness, Tick-Borne Relapsing Fever, and tularemia.

9 REFERENCES

[1] Laschewski, G., G. Jendritzky, 2002. Effects of the Thermal Environment on Human Health: An Investigation of 30 Years of Daily Mortality Data from SW Germany. Climate Research, 21, 91-103.

[2] Kysely, J., L. Pokorna, J. Kyncl, B. Kriz, 2009. Excess Cardiovascular Mortality Associated with Cold Spells in the Czech Republic. BMC Public Health, 9, 19.

[3] Ma, W., C. Yang, C. Chu, T. Li, J. Tan, H. Kan, 2013. The Impact of the 2008 Cold Spell on Mortality in Shanghai, China. International Journal of Biometeorology, 57, 179-184.

[4] Mölders, N., 2011. Land-Use and Land-Cover Changes: Impact on Climate and Air Quality. Vol. 44. Springer Science & Business Media. 193.

[5] Leiter, U., U. Keim, C. Garbe, 2020. Epidemiology of Skin Cancer: Update 2019. Sunlight, Vitamin D and Skin Cancer, 123-139.

[6] Gilligan, S.P., 2008. Identifying Communities of Practice through Cord Impressed Ceramics at the Early Late Prehistoric Sites of Thomas/Luckey and Chenango Point in Central New York. State University of New York at Binghamton.

[7] Lu, X., R.C. Clarke, 1995. The Cultivation and Use of Hemp (Cannabis Sativa L.) in Ancient China. Journal of the International Hemp Association, 2, 26-30.

[8] Skoglund, P., 2008. Stone Ships: Continuity and Change in Scandinavian Prehistory. World Archaeology, 40, 390-406.

[9] Sundbø, A., 2001. Setesdal Sweaters the History of the Norwegian Lice Pattern. Torridal Tweed. 155 pp.

[10] Özdemir, H., 2017. Thermal Comfort Properties of Clothing Fabrics Woven with Polyester/Cotton Blend Yarns. AUTEX Research Journal, 17, 135-141.

[11] Döbler, H., 1972. Kultur- und Sittengeschichte der Welt, Kleidung, Mode, Schmuck, Bertelsmann Verlag, 358p.

[12] Mohiuddin, A., M.K. Saha, M.S. Hossian, A. Ferdoushi, 2014. Usefulness of Banana (Musa Paradisiaca) Wastes in Manufacturing of Bio-Products: A Review. The Agriculturists, 12, 148-158.

[13] Gelbke, H.-P., T. Göen, M. Mäurer, S.I. Sulsky, 2009. A Review of Health Effects of Carbon Disulfide in Viscose Industry and a

Proposal for an Occupational Exposure Limit. Critical Reviews in Toxicology, 39, 1-126.

[14] Oglakcioglu, N., P. Celik, T.B. Ute, A. Marmarali, H. Kadoglu, 2009. Thermal Comfort Properties of Angora Rabbit/Cotton Fiber Blended Knitted Fabrics. Textile Research Journal, 79, 888-894.

[15] Boughattas, A., S. Benltoufa, M. Azeem, L. Hes, F. Fayala, 2018. Thermo-Physiological Comfort of Brushed Woven Fabrics. Fibres and Textiles (Vlákna a Textil), 25, 7-12.

[16] Mölders, N., G. Kramm, 2014. Lectures in Meteorology. Heidelberg: Springer. 591p.

[17] Kramm, G., R. Dlugi, 2011. Scrutinizing the Atmospheric Greenhouse Effect and Its Climatic Impact. Natural Science, 3, 971-998.

[18] Mölders, N., 2019. Outdoor Universal Thermal Comfort Index Climatology for Alaska. Atmospheric and Climate Sciences, 9, 558-582.

[19] Skin Cancer Foundation, 2023. Cited 2023 6/12/2023. Available from: https://www.skincancer.org/.

[20] Sarkar, A.K., 2004. An Evaluation of UV Protection Imparted by Cotton Fabrics Dyed with Natural Colorants. BMC Dermatology, 4, 1-8.

[21] Syed, U., 2010. The Influence of Woven Fabrics Structures on the Continuous Dyeing of Lyocell Fabrics with Reactive Dyes. Doctoral dissertation. Heriot-Watt University Scottish Border Campus Galashiels, p. 268.

[22] Dubrovski, P.D., D. Golob, 2009. Effects of Woven Fabric Construction and Color on Ultraviolet Protection. Textile Research Journal, 79, 351-359.

[23] Kocić, A., M. Bizjak, D. Popović, G.B. Poparić, S.B. Stanković, 2019. UV Protection Afforded by Textile Fabrics Made of Natural and Regenerated Cellulose Fibres. Journal of Cleaner Production, 228, 1229-1237.

[24] Özkan, E.T., B.M. Kaplangiray, 2019. Investigating Thermo-physiological Comfort Properties of Polyester Knitted Fabrics. Journal of Textile Engineering & Fashion Technology, 5, 50-56.

[25] Vimal, J.T., R. Murugan, V. Subramaniam, 2016. Effect of Weave Parameters on the Air Resistance of Woven Fabrics. Fibres & Textiles in Eastern Europe, 24, 67-72.

[26] Bedek, G., F. Salaün, Z. Martinkovska, E. Devaux, D. Dupont, 2011. Evaluation of Thermal and Moisture Management Properties on Knitted Fabrics and Comparison with a Physiological Model in Warm Conditions. Applied Ergonomics, 42, 792-800.

[27] Matusiak, M., K. Sikorski, 2011. Influence of the Structure of Woven Fabrics on Their Thermal Insulation Properties. Fibres & Textiles in Eastern Europe, 88, 46-53.

[28] Elnashar, E.A., 2017. Volume Porosity and Air Permeability in Knitting Fabrics. International Journal of Research in Advanced Engineering and Technology, 3, 75-80.

[29] Karthikeyan, G., G. Nalankilli, O. Shanmugasundaram, C. Prakash, 2016. Thermal Comfort Properties of Bamboo Tencel Knitted Fabrics. International Journal of Clothing Science and Technology. 10.1108/IJCST-08-2015-0086.

[30] Stanković, S.B., D. Popović, G.B. Poparić, 2008. Thermal Properties of Textile Fabrics Made of Natural and Regenerated Cellulose Fibers. Polymer Testing, 27, 41-48.

[31] Prakash, C., G. Ramakrishnan, 2014. Study of Thermal Properties of Bamboo/Cotton Blended Single Jersey Knitted Fabrics. Arabian Journal for Science and Engineering, 39, 2289-2294.

[32] Prakash, C., G. Ramakrishnan, C.V. Koushik, 2012. A Study of the Thermal Properties of Single Jersey Fabrics of Cotton, Bamboo and Cotton/Bamboo Blended-Yarn Vis-a-Vis Bamboo Fibre Presence and Yarn Count. Journal of Thermal Analysis and Calorimetry, 110, 1173-1177.

[33] Yavaşcaoğlu, A., E. Recep, S. Gülcan, 2018. Effects of Usage Acrylic Yarn on Thermal Comfort and Moisture Management Properties of Woven Shirting Fabrics. Textile and Apparel, 28, 141-150.

[34] Özdil, N., A. Marmarali, S.D. Kretzschmar, 2007. Effect of Yarn Properties on Thermal Comfort of Knitted Fabrics. International Journal of Thermal Sciences, 46, 1318-1322.

[35] Ahmad, S., F. Ahmad, A. Afzal, A. Rasheed, M. Mohsin, N. Ahmad, 2015. Effect of Weave Structure on Thermo-Physiological Properties of Cotton Fabrics. AUTEX Research Journal, 15, 30-34.

[36] Majumdar, A., S. Mukhopadhyay, R. Yadav, 2010. Thermal Properties of Knitted Fabrics Made from Cotton and Regenerated

Bamboo Cellulosic Fibres. International Journal of Thermal Sciences, 49, 2042-2048.

[37] Stanković, S.B., M. Novaković, D.M. Popović, G.B. Poparić, M. Bizjak, 2019. Novel Engineering Approach to Optimization of Thermal Comfort Properties of Hemp Containing Textiles. The Journal of The Textile Institute, 110, 1271-1279.

[38] Bhattacharjee, D., V. Kothari, 2009. Heat Transfer through Woven Textiles. International Journal of Heat and Mass Transfer, 52, 2155-2160.

[39] Shawky, M., H. Darwish, 2016. Comfort Properties of Some Cellulosic Fabrics. International Journal of ChemTech Research, 9, 266-277.

[40] Yildiz, Z., N.O. Ceviz, V. Dal, E. Akcagun, A. Yilmaz, M. Atmaca, A.B. Kurtulus, 2011. Investigation of Physical Properties of Woolen Fabrics on Thermal Comfort. Power Energy and Society, 162, 0-25.

[41] Cubric, I.S., Z. Skenderi, G. Havenith, 2013. Impact of Raw Material, Yarn and Fabric Parameters, and Finishing on Water Vapor Resistance. Textile Research Journal, 83, 1215-1228.

[42] Havlová, M., 2020. Air Permeability, Water Vapour Permeability and Selected Structural Parameters of Woven Fabrics. Fibres and Textiles, 27, 12-18.

[43] Özkan, E.T., M. Binnaz, 2015. Thermophysiological Comfort Properties of Different Knitted Fabrics Used in Cycling Clothes. Textile Research Journal, 85, 62-70.

[44] Elnashar, E., G. Bashkova, 2014. Comfort of Thermal Insulation of Bulky Woven Fabrics for Clothes. In: International Conference on Technics, Technologies and Education ICTTE, 2, 329-344.

[45] Birrfelder, P., M. Dorrestijn, C. Roth, R.M. Rossi, 2013. Effect of Fiber Count and Knit Structure on Intra- and Inter-Yarn Transport of Liquid Water. Textile Research Journal, 83, 1477-1488.

[46] Babu, B.S., P. Senthilkumar, M. Senthilkumar, 2015. Effect of Yarn Linear Density on Moisture Management Characteristics of Cotton/Polypropylene Double Layer Knitted Fabrics. Industria Textila, 66, 123-130.

[47] Premalatha, C., M.R. Kumar, S. Udaya Krithika, C. Prakash, 2023. Analysis of Thermal Comfort Properties of Tri-Layer Knitted Fabrics. Journal of Natural Fibers, 20, 2164107.

[48] Lei, M., Y. Li, Y. Liu, Y. Ma, L. Cheng, Y. Hu, 2020. Effect of Weaving Structures on the Water Wicking–Evaporating Behavior of Woven Fabrics. Polymers, 12, 422.

[49] Saravanan, D., 2007. UV Protection Textile Materials. AUTEX Research Journal, 7, 53-64.

[50] Kan, C.W., 2014. A Study on Ultraviolet Protection of 100% Cotton Knitted Fabric: Effect of Fabric Parameters. The Scientific World Journal, 2014, Article ID 506049, 10 p.

[51] Ibrahim, S.F., D.M. Essa, E.M. Osman, 2016. Spectral Evaluation Studies on Titanium Dioxide Nano-Particles and Their Add-Mixtures on Textile Fabrics. Egyptian Journal of Chemistry, 59, 1069-1093.

[52] Zakaria, M.E.H. Nizam, M.H.A. Mamun, M.A. Yousuf, R. Ali, L. Rahman, M.R. Miah, 2017. Dyeing of S/J Cotton Knit Fabric with Natural Dye Extracts from Green Walnut Shells: Assessment of Mordanting Effect on Fastness Properties. Journal of Textile Science and Technology, 3, 17-30.

[53] Rungruangkitkrai, N., N. Tubtimthai, A. Cholachatpinyo, R. Mongkholrattanasit, 2012. UV Protection Properties of Wool Fabric Dyed with Eucalyptus Leaf Extract by the Padding Techniques. In: RMUTP International Conference: Textiles and Fashion, Bangkok. 2012.

[54] Singh, M.K., A. Singh, 2013. Ultraviolet Protection by Fabric Engineering. Journal of Textiles, 2013 (ID 579129), 6 p.

[55] Mongkholrattanasit, R., J. Kryštůfek, J. Wiener, M. Viková, 2011. UV Protection Properties of Silk Fabric Dyed with Eucalyptus Leaf Extract. The Journal of the Textile Institute, 102, 272-279.

[56] Rungruangkitkrai, N., R. Mongkholrattanasit, P. Phoophat, N. Chartvivatpornchai, S. Sirimungkararat, K. Wongkasem, P. Tuntariyanond, N. Nithithongsakol, R. Chollakup, 2020. UV-Protection Property of Eri Silk Fabric Dyed with Natural Dyes for Eco-Friendly Textiles. In: IOP Conference Series: Materials Science and Engineering 2020, 773, 012027, 4 p.

[57] Badr, A.A., A. Hassanin, M. Moursey, 2016. Influence of Tencel/Cotton Blends on Knitted Fabric Performance. Alexandria Engineering Journal, 55, 2439-2447.

[58] Gawish, S., H. Helmy, A. Ramadan, R. Farouk, H. Mashaly, 2016. UV Protection Properties of Cotton, Wool, Silk and Nylon Fabrics

Dyed with Red Onion Peel, Madder and Chamomile Extracts. Journal of Textile Science & Engineering, 6, 1-13.

[59] Sohaib Anas, M., A. Abbas, Z. Azam, Z. Tariq, Z. Gul, M.E. Sarwar, 2022. Ultraviolet Protection Factor Evaluation of Comfort Oriented Two-Yarn Fleece Fabrics. Journal of Engineered Fibers and Fabrics, 17, 1-13.

[60] Akgun, M., B. Becerir, H.R. Alpay, 2010. Assessing the Relationship among Fabric Constructional Parameters, Fractional Reflectances and Cover Factors of Polyester Fabrics by Experimental and Mathematical Methods. Fibers and Polymers, 11, 291-302.

[61] Hou, X., X. Chen, Y. Cheng, H. Xu, L. Chen, Y. Yang, 2013. Dyeing and UV-Protection Properties of Water Extracts from Orange Peel. Journal of Cleaner Production, 52, 410-419.

[62] Hussain, T., A. Nazir, R. Masood, 2015. Liquid Moisture Management in Knitted Textiles – A Review. In: 3rd International Conference on Value Addition & Innovation in Textiles (Covitex-2015), 27-28 March 2015, National Textile University Faisalabad Pakistan. Conference Proceedings, 15-26.

[63] Kandi, I., 2013. High Performance & Colour Strength Behavior of Bamboo/Polyester Blended Woven Fabrics. International Journal of Innovative Research in Science, Engineering and Technology, 2, 7620-7629.

[64] Savvidis, G., V. Karanikas, M. Zarkogianni, I. Eleftheriadis, N. Nikolaidis, E. Tsatsaroni, 2017. Screen-Printing of Cotton with Natural Pigments: Evaluation of Color and Fastness Properties of the Prints. Journal of Natural Fibers, 14, 326-334.

[65] Ghosh, S., P. Bajaj, V. Kothari, 2003. Effect of Dyes and Finishes on UV Protection of Jute/Cotton Fabrics. Indian Journal of Fibre & Textile Research, 28, 431-436.

[66] Farrell, M.J., G.A. Vandezande, 2015. Ultraviolet Protection and Light Transmission Properties of a Cotton/Spandex Fabric Treated with TiO_2 Extender Pigments. AATCC Journal of Research, 2, 20-25.

[67] Rood, E.S., 1921. Thermal Conductivity of Some Wearing Materials. Physical Review, 18, 356.

[68] Yavuzkasap Ayakta, D., E. Öner, S. Celikkiran, 2019. A Research on the Mechanical Properties of Worsted Fabrics Made of High Tenacity Polyamide. Fibers and Polymers, 20, 2426-2432.

[69] Xu, X., T.P. Rioux, M.P. Castellani, 2022. The Specific Heat of the Human Body Is Lower Than Previously Believed: The Journal Temperature Toolbox. Temperature, 1-5.

[70] Oğlakcioğlu, N., A. Marmarali, 2007. Thermal Comfort Properties of Some Knitted Structures. Fibres & Textiles in Eastern Europe, 15, 64-65.

[71] Marmarali, A., H. Kadoglu, N. Oglakcioglu, P. Celik, M. Blaga, M. Ursache, C. Loghin, 2009. Thermal Comfort Properties of Some New Yarns Generation Knitted Fabrics. In: AUTEX 2009 World Textile Conference, 26-28 May, 2009 Izmir, Turkey. 7 p.

[72] Kumari, A., K. Khurana, 2016. Research Article Regenerated Cellulose-Based Denim Fabric for Tropical Regions: An Analytical Study on Making Denim Comfortable. Journal of Textiles, 2016 (4614168), 10 p.

[73] Atalie, D., A.F. Tesema, G.K. Rotich, 2018. Effect of Weft Yarn Twist Level on Thermal Comfort of 100 Per Cent Cotton Woven Fabrics. Research Journal of Textile and Apparel, 22, 180-194.

[74] Hashan, M., K.F. Hasan, F. Khandaker, K. Karmaker, Z. Deng, M.J. Zilani, 2017. Functional Properties Improvement of Socks Items Using Different Types of Yarn. International Journal of Textile Science, 6, 34-42.

[75] Ertekin, G., A. Marmarali, 2011. Heat, Air and Water Vapor Transfer Properties of Circular Knitted Spacer Fabrics. Textile and Apparel, 21, 369-373.

[76] Nagy, B., S.G. Nehme, D. Szagri, 2015. Thermal Properties and Modeling of Fiber Reinforced Concretes. Energy Procedia, 78, 2742-2747.

[77] Adnan, M., J. Moses, 2020. A Study on the Thermophysiological and Tactile Comfort Properties of Silk/Lyocell Blended Fabrics. Matéria, Rio de Janeiro, 25, 9 p.

[78] Subburaayasaran, A., S.K. Sampath Kumar, D. Vasanth Kumar, T. Ramachandran, C. Prakash, H. Vijayakumar, 2022. Comparative Studies on Thermal Comfort Properties of Eri Silk, Wool/Eri Silk, Cotton, and Micro-Denier Acrylic Double-Layered Knitted Fabrics. Journal of Natural Fibers. 10, 1-9.

[79] Onofrei, E., A. Maria Rocha, A. Catarino, 2012. Investigating the Effect of Moisture on the Thermal Comfort Properties of Functional Elastic Fabrics. Journal of Industrial Textiles, 42, 34-51.

[80] Braga, I., M.J. Abreu, M. Oliveira, 2018. The Importance of a Comparative Thermophysiological Study in Female Jeans. In: IOP Conference Series: Materials Science and Engineering, 459, Aegean International Textile and Advanced Engineering Conference (AITAE 2018), 5–7 September 2018, Lesvos, Greece, 012034, 8 p.

[81] Akcagun, E., M. Bogusławska-Bączek, L. Hes, 2019. Thermal Insulation and Thermal Contact Properties of Wool and Wool/PES Fabrics in Wet State. Journal of Natural Fibers, 16, 199-208.

[82] Mölders, N., 2023. Inventory of the Thermo-Physiological Behavior of Fabrics – A Review. Journal of Textile Science and Technology, 9, 127-150.

[83] Manyam, S., A. Padma, D. Anitha, 2013. Comfort Properties of Ecologically Friendly Sisal Union Fabrics. Enzyme, 4, 20-25.

[84] Sharma, A., S. Pant, 2013. Studies on Camel Hair-Merino Wool Blended Knitted Fabrics. Indian Journal of Fibre & Textile Research, 38, 317-319.

[85] Dawoud, A.A., 2016. Effect of Weave Structure on Comfort Properties of Microfiber Polyester Woven Fabrics. International Journal of Engineering Sciences & Management Research, 3, 26-33.

[86] Prakash, C., G. Ramakrishnan, C. Koushik, 2013. Effect of Blend Proportion on Moisture Management Characteristics of Bamboo/Cotton Knitted Fabrics. The Journal of The Textile Institute, 104, 1320-1326.

[87] Shaker, K., M. Umair, M. Jabbar, D.M. Baitab, Y. Nawab, A. Afzal, S. Ahmad, 2019. Effect of Fabric Structural Design on the Thermal Properties of Woven Fabrics. Thermal Science, 23, 3059-3066.

[88] Atalie, D., P. Tesinova, M. Tadesse, E. Ferede, I. Dulgheriu, E. Loghin, 2021. Materials, 14, 6863, 12 p.

[89] Özgen, B., S. Altaş, 2014. The Investigation of Thermal Comfort, Moisture Management and Handle Properties of Knitted Fabrics Made of Various Fibres. Textile and Apparel, 24, 272-278.

[90] Rani, K., L. Jajpura, B. Behera, 2019. Comfort Behavior of Unconventional Natural Fiber-Based Union Fabrics. Journal of Textile Science and Technology, 5, 96442, 9 p.

[91] Islam, M.R., U.K. Fatema, 2021. A Comparative Analysis of Polyester Fabric Properties between Dyed with Indigo and with

Disperse Dyes. Journal of Textile Science and Technology, 7, 77-90.

[92] Sheikh, M.R., M.K. Hossain, M.M.R. Khan, 2022. Influence of Lycra Content and Stitch Length on the Dimensional and Physical Characteristics of Lycra Back Plaited Cotton (LBPC) Single Jersey Fabrics. Journal of Textile Science and Technology, 8, 203-220.

[93] Saleh, S.M., K. El-Badry, 2012. Dyeing of Cotton/Nylon Blended Fabric to a Solid Shade in One Bath. Global Journal of Science Frontier Research, 12 (7-D), 8 p.

[94] Rosace, G., E. Guido, C. Colleoni, G. Barigozzi, 2016. Influence of Textile Structure and Silica-Based Finishing on Thermal Insulation Properties of Cotton Fabrics. International Journal of Polymer Science, 2016, 1726475, 10 p.

[95] Pant, S., R. Jain, 2014. Comfort and Mechanical Properties of Cotton and Cotton Blended Knitted Khadi Fabrics. Studies on Home and Community Science, 8, 69-74.

[96] Lakhchaura, P., M. Gahlot, 2021. A Study on Comfort Properties of Oak Tasar Silk Waste and Acrylic Blended Fabrics. Journal of Applied and Natural Science, 13, 476-481.

[97] Verma, N., N. Grewal, S. Bains, 2016. Evaluation of Comfort and Handle Behavior of Mulberry Silk Waste/Wool Blended Fabrics for End Use. Journal of Natural Fibers, 13, 277-288.

[98] Hawas, H., 2021. A Study on Comfort Properties of Banana/Cotton Blend Woven Fabrics. Journal of Design Sciences and Applied Arts, 2, 96-105.

[99] Mangat, M.M., T. Hussain, V. Bajzik, 2012. Impact of Different Weft Materials and Washing Treatments on Moisture Management Characteristics of Denim. Journal of Engineered Fibers and Fabrics, 7, 38-49.

[100] Tashkandi, S., 2014. An Investigation of Thermal Comfort Properties of Abaya under Heat Stress. Doctoral Dissertation, RMIT University, Melborne Australia.

[101] Chakroun, M.G., S. Benltoufa, F. Fayala, 2021. The Effect of Fabric's Structure on the Breathability and the Drying Rate Properties. Communications in Development and Assembling of Textile Products, 2, 61-69.

[102] Das, B., A. Das, V. Kothari, R. Fanguiero, M.D. Araujo, 2009. Moisture Flow through Blended Fabrics–Effect of Hydrophilicity. Journal of Engineered Fibers and Fabrics, 4, 20-28.

[103] Karaca, E., N. Kahraman, S. Omeroglu, B. Becerir, 2012. Effects of Fiber Cross Sectional Shape and Weave Pattern on Thermal Comfort Properties of Polyester Woven Fabrics. Fibres & Textiles in Eastern Europe, 92, 67-72.

[104] Hasan, M.K., J. Kayumov, G. Zhu, M. Khatun, A. Nur, X. Ding, 2019. An Experimental Investigation to Examine the Wicking Properties of Silk Fabrics. Journal of Textile Science and Technology, 5, 108-124.

[105] Kumar, P., S.K. Sinha, S. Ghosh, 2015. Moisture Management Behaviour of Modified Polyester Wool Fabrics. Fashion and Textiles, 2, 1-17.

[106] Barsha, U.H., M. Uddin, M. Hasan, S. Chowdhury, 2018. Investigation on Physical Properties of Jute Cotton (Jutton) Blended Fabric. International Journal of Textile Fashion Technology, 8, 1-8.

[107] Chen, Q., K.-P.M. Tang, P. Ma, G. Jiang, C. Xu, 2017. Thermophysiological Comfort Properties of Polyester Weft-Knitted Fabrics for Sports T-Shirt. The Journal of the Textile Institute, 108, 1421-1429.

[108] Venkatraman, P., 2015. Fabric Properties and Their Characteristics. Materials and Technology for Sportswear and Performance Apparel, 18, 53-86.

[109] Vidhya, M., K. Parveen Banu, V. Kumar D, C. Prakash, V. Subramaniyam, 2022. Moisture Management Properties of Single Jersey Knitted Fabrics Made from Regenerated Bamboo: Polyester Core-Spun Yarns. Journal of Natural Fibers, 19, 15377-15391.

[110] Matusiak, M., 2019. Moisture Management Properties of Seersucker Woven Fabrics of Different Structure. Fibres & Textiles in Eastern Europe, 3, 43-50.

[111] Singh, M.K., D.B. Shakyawar, P. Bajpai, A.K.S. Gangwar, 2023. Pashmina Fiber Blended Woven Fabrics for High Functional Performances. Journal of Natural Fibers, 20, 1-12.

[112] Jin Lee, J., D. Sun Ji, 2017. Evaluation of Liquid Moisture Management Properties on Hemp Woven Fabrics Treated with Liquid Ammonia. Textile Research Journal, 87, 1752-1764.

[113] Marolleau, A., F. Salaün, D. Dupont, H. Gidik, S. Ducept, 2017. Influence of Textile Physical Properties and Thermo-Hydric Behaviour on Comfort. Journal of Ergonomics, 7, 1000223, 9 p.

[114] Öner, E., A. Okur, 2015. Thermophysiological Comfort Properties of Selected Knitted Fabrics and Design of T-Shirts. The Journal of The Textile Institute, 106, 1403-1414.

[115] Okur, N., 2021. Thermo-Physiological and Handle-Related Comfort Properties of Hemp and Flax Blended Denim Fabrics. Journal of Natural Fibers, 19, 1-14.

[116] Saricam, C., 2022. The Comfort Properties of Hemp and Flax Blended Denim Fabrics with Common Industrial Washing Treatments. Textile Research Journal, 92, 3164-3178.

[117] Lei, Z., 2019. Review of Application of Thermal Manikin in Evaluation on Thermal and Moisture Comfort of Clothing. Journal of Engineered Fibers and Fabrics, 14, 1-10.

[118] ASHARE, 2001. American Society of Heating, Refrigerating, and Air-Conditioning Engineers, USA.

[119] Mölders, N., 2016. How to Dress for Success in Midlife. San Bernadino, CA, USA: Creative Space. 96.

[120] Zhang, X., X. Chao, L. Lou, J. Fan, Q. Chen, B. Li, L. Ye, D. Shou, 2021. Personal Thermal Management by Thermally Conductive Composites: A Review. Composites Communications, 23, 100595, 12 p.

[121] Farooq, A.S., P. Zhang, 2021. Fundamentals, Materials and Strategies for Personal Thermal Management by Next-Generation Textiles. Composites Part A: Applied Science and Manufacturing, 142, 106249, 11p.

[122] Skutecki, R., R. Jalali, E. Dragańska, I. Cymes, J. Romaszko, K. Glińska-Lewczuk, 2019. UTCI as a Bio-Meteorological Tool in the Assessment of Cold-Induced Stress as a Risk Factor for Hypertension. Science of The Total Environment, 688, 970-975.

[123] Mercer, J.B., 2003. Cold—An Underrated Risk Factor for Health. Environmental Research, 92, 8-13.

[124] Salata, F., I. Golasi, R. De Lieto Vollaro, A. De Lieto Vollaro, 2016. Outdoor Thermal Comfort in the Mediterranean Area. A Transversal Study in Rome, Italy. Building and Environment, 96, 46-61.

[125] Jendritzky, G., W. Nubler, 1981. A Model Analyzing the Urban Thermal Environment in Physiologically Significant Terms. Archives of Meteorology and Geophysical Bioclimatology, 29, 313-326.

[126] Pickup, J., R. De Dear, 2000. WMO, WCASP, Geneva. 279-283.

[127] Blazejczyk, K., Y. Epstein, G. Jendritzky, H. Staiger, B. Tinz, 2012. Comparison of UTCI to Selected Thermal Indices. International Journal of Biometeorology, 56, 515-535.

[128] Thorsson, S., M. Lindqvist, S. Lindqvist, 2004. Thermal Bioclimatic Conditions and Patterns of Behaviour in an Urban Park in Göteborg, Sweden. International Journal of Biometeorology, 48, 149-156.

[129] Holst, J., H. Mayer, 2011. Impacts of Street Design Parameters on Human-Biometeorological Variables. Meteorologische Zeitschrift, 20, 541-552.

[130] Golasi, I., F. Salata, E. De Lieto Vollaro, M. Coppi, 2018. Complying with the Demand of Standardization in Outdoor Thermal Comfort: A First Approach to the Global Outdoor Comfort Index (GOCI). Building and Environment, 130, 104-119.

[131] Havenith, G., D. Fiala, K. Błazejczyk, M. Richards, P. Bröde, I. Holmér, H. Rintamaki, Y. Benshabat, G. Jendritzky, 2012. The UTCI-Clothing Model. International Journal of Biometeorology, 56, 461-470.

[132] Fiala, D., G. Havenith, P. Bröde, B. Kampmann, G. Jendritzky, 2012. UTCI-Fiala Multi-Node Model of Human Heat Transfer and Temperature Regulation. International Journal of Biometeorology, 56, 429-441.

[133] Bröde, P., D. Fiala, K. Błażejczyk, I. Holmér, G. Jendritzky, B. Kampmann, B. Tinz, G. Havenith, 2012. Deriving the Operational Procedure for the Universal Thermal Climate Index (UTCI). International Journal of Biometeorology, 56, 481-494.

[134] Sormunen, E., 2009. Repetitive Work in the Cold: Work Ability, Musculoskeletal Symptoms and Thermal and Neuromuscular Responses in Food Industry Workers. Academic dissertation, Faculty of Medicine of the University of Oulu, Universitatis Oulensis. p. 108.

[135] Mölders, N., 2019. Outdoor Universal Thermal Comfort Index Climatology for Alaska. Atmospheric and Climate Sciences, 9, 558-582.

[136] Huang, J., J.G. Cedeño-Laurent, J.D. Spengler, 2014. Citycomfort+: A Simulation-Based Method for Predicting Mean Radiant Temperature in Dense Urban Areas. Building and Environment, 80, 84-95.

[137] Olusegun, H.D., 2006. Investigation into Effects of Clothing in Saharan Regions. Journal of the Energy Institute, 79, 172-179.

[138] Seinfeld, J.H., S.N. Pandis, 1997. Atmospheric Chemistry and Physics, from Air Pollution to Climate Change. John Wiley & Sons. 1119 p.

[139] Kramm, G., R. Dlugi, N. Mölders, 2004. On the Vertically Averaged Balance Equation of Atmospheric Trace Constituents. Meteorology and Atmospheric Physics, 86, 121-141.

[140] Kramm, G., N. Mölders, M. Berger, R. Dlugi, 2022. On the Solar Climate of the Moon and the Resulting Surface Temperature Distribution. Natural Science, 14, 386-420.

[141] World Meteorological Organization, 1998. Climate Statistics, WMO, Genova.

[142] Köppen, W., 1884. Die Wärmezonen der Erde, nach der Dauer der heissen, gemässigten und kalten Zeit und nach der Wirkung der Wärme auf die organische Welt betrachtet. Meteorologische Zeitschrift, 1, 215-226.

[143] Köppen, W., 1900. Versuch einer Klassifikation der Klimate, vorzugsweise nach ihren Beziehungen zur Pflanzenwelt. Geographische Zeitschrift 6, 593–611, 657–679.

[144] Geiger, R., 1961. Überarbeitete Neuausgabe von Köppen–Geiger/Klima der Erde. Klett Perthes, Gotha.

[145] Kottek, M., J. Grieser, C. Beck, B. Rudolf, F. Rubel, 2006. World Map of the Köppen-Geiger Climate Classification Updated. Meteorologische Zeitschrift, 15, 259-263.

[146] Peel, M.C., B.L. Finlayson, T.A. McMahon, 2007. Updated World Map of the Köppen-Geiger Climate Classification. Hydrology Earth System Sciences, 11, 1633-1644.

[147] Rubel, F., K. Brugger, K. Haslinger, I. Auer, 2017. The Climate of the European Alps: Shift of Very High-Resolution Köppen-Geiger Climate Zones 1800-2100. Meteorologische Zeitschrift, 26, 115-125.

[148] Beck, H.E., N.E. Zimmermann, T.R. McVicar, N. Vergopolan, A. Berg, E.F. Wood, 2018. Present and Future Köppen-Geiger Climate Classification Maps at 1-Km Resolution. Scientific Data, 5, 180214.

[149] Skenderi, Z., I. Salopek Čubrić, B. Mijović, 2011. The Comparison of Human Perception to the Measured Sweat Transfer Rate. Collegium Antropologicum, 35, 433-437.

[150] Anisimov, O.A., F.E. Nelson, 1996. Permafrost Distribution in the Northern Hemisphere under Scenarios of Climatic Change. Global and Planetary Change, 14, 59-72.

[151] Błażejczyk, K., J. Baranowski, G. Jendritzky, A. Błażejczyk, P. Bröde, D. Fiala, 2015. Regional Features of the Bioclimate of Central and Southern Europe against the Background of the Köppen-Geiger Climate Classification. Geographia Polonica, 88, 439-453.

[152] Gommershtadt, O., P. Konstantinov, M. Varentsov, A. Baklanov, 2020. Modeling Technology for Assessment of Summer Thermal Comfort Conditions of Arctic City on Microscale: Application for City of Apatity. In: Green Technologies and Infrastructure to Enhance Urban Ecosystem Services: Proceedings of the Smart and Sustainable Cities Conference 2018. Springer International Publishing, 66-75.

[153] Revich, B., D. Shaposhnikov, 2022. The Influence of Heat and Cold Waves on Mortality in Russian Subarctic Cities with Varying Climates. International Journal of Biometeorology, 66, 2501-2515.

[154] Bhatia, D., A. Sharma, U. Malhotra, 2014. Recycled Fibers: An Overview. International Journal of Fiber and Textile Research, 4, 77-82.

[155] Suganthi, T., P. Senthilkumar, V. Dipika, 2017. Thermal Comfort Properties of a Bi-Layer Knitted Fabric Structure for Volleyball Sportswear. Fibres & Textiles in Eastern Europe, 25, 75-80.

[156] Çeven, E.K., G.K. Günaydin, 2018. Investigation of Moisture Management and Air Permeability Properties of Fabrics with Linen and Linen-Polyester Blend Yarns. Fibres & Textiles in Eastern Europe, 4, 39-47.